Contributions to Economics

The series *Contributions to Economics* provides an outlet for innovative research in all areas of economics. Books published in the series are primarily monographs and multiple author works that present new research results on a clearly defined topic, but contributed volumes and conference proceedings are also considered. All books are published in print and ebook and disseminated and promoted globally. The series and the volumes published in it are indexed by Scopus and ISI (selected volumes).

Eskil Ullberg

Intellectual Property Statistics

Measuring Framework for Standards and Trade in Ideas

Foreword by Robert Koopman

Springer

Eskil Ullberg
Department of Economics
College of Humanities and Social
Sciences, George Mason University
Arlington, VA, USA

Foreword by
Robert Koopman
School of International Service
American University
Washington, DC, USA

ISSN 1431-1933 ISSN 2197-7178 (electronic)
Contributions to Economics
ISBN 978-3-031-36385-6 ISBN 978-3-031-36386-3 (eBook)
https://doi.org/10.1007/978-3-031-36386-3

This Springer imprint is published by the registered company Springer Nature Switzerland AG
The registered company address is: Gewerbestrasse 11, 6330 Cham, Switzerland

Paper in this product is recyclable.

To the international policy makers, economists, statisticians, and accountants in the world who wondered why their statistics never measured up to the economic system they each and every day observed around them, and to the inventors, business managers, investors, and university-institute researchers who always knew that their work with developing, trading, and using ideas in intellectual property form had economic impact but was never fully recognized neither in their private accounting nor in the public statistics.

Foreword

In this book, *Intellectual Property Statistics*, Prof. Ullberg has undertaken a Herculean task – to lay out a paradigm for the collection of intellectual property statistics with the aim to ensure that the developing and evolving market of trade in ideas has the information and data necessary to function well.

The task is Herculean because not only do we not have comprehensive data on trade in IP, but first one needs to establish the appropriate conceptual and statistical frameworks so that statisticians can collect, organize, analyze, and report consistent and high quality information.

In the world of global economic statistics, such frameworks are often the work of large statistical organizations such as the US Bureau of Economic Analysis, Statistics Canada, EuroStat, or UN Statistics. Years of formal meetings of leading technical experts, supported by deep government and international organization budgets, will work to lay out the needed concepts and procedures.

In this volume, Prof. Ullberg has managed to start this important conversation with some funding from the Government of Sweden and through his efforts to bring a small group of willing international experts together, and finally to test some of these ideas with a group of developing countries to explore how they might work.

As with any foray into a new and complex area, this volume should be viewed as a starting point, a work in progress, but an important one that could very well influence the development of this important set of data on trade in ideas. At a time when global issues such as climate change and global health pandemics require both new ideas and the spread of those ideas widely to help ensure both economic growth and continued global economic convergence, data that help us monitor and evaluate what is happening in trade in ideas will be extremely valuable.

Robert Koopman

Hurst Senior Professorial Lecturer,
School of International Service,
American University, Washington, DC,
USA

Former Chief Economist and Director of
the Economic Research and Statistics
Division, World Trade Organization,
Geneva, Switzerland

Preface

The problem addressed in this book is the development of new Intellectual Property (IP) statistics for a world economic system that is *idea based*, i.e., increasingly depends on *trade in ideas (IP)*. This is a topic which is covered incompletely at best in today's balance of payments, financial accounting, and statistical statements. The current principles are based on trading "things in possession," physical goods which are moved across borders and serviced by a range of services from guarantees, finance, and technical support to management of performance and risk sharing arrangements. The proposed IP statistics is instead based on the principle of "things in action" being the key economic activity of the twenty-first century, focusing on the trading of IP rights, starting with patents. Patents are granted for new technology and are transferrable and licensable rights, thus promising specialization through trade and increased technology growth and productivity growth which is at the heart of the economic growth.

Failing to analyze the economic system from this action-rights angle may lead to an incomplete understanding of this dynamic economic system centered on inventions and human creativity – especially for developing nations who now also have high levels of human capital formation – and to wrong business and economic policies, not properly leveraging the rarest flowers of human ideas.

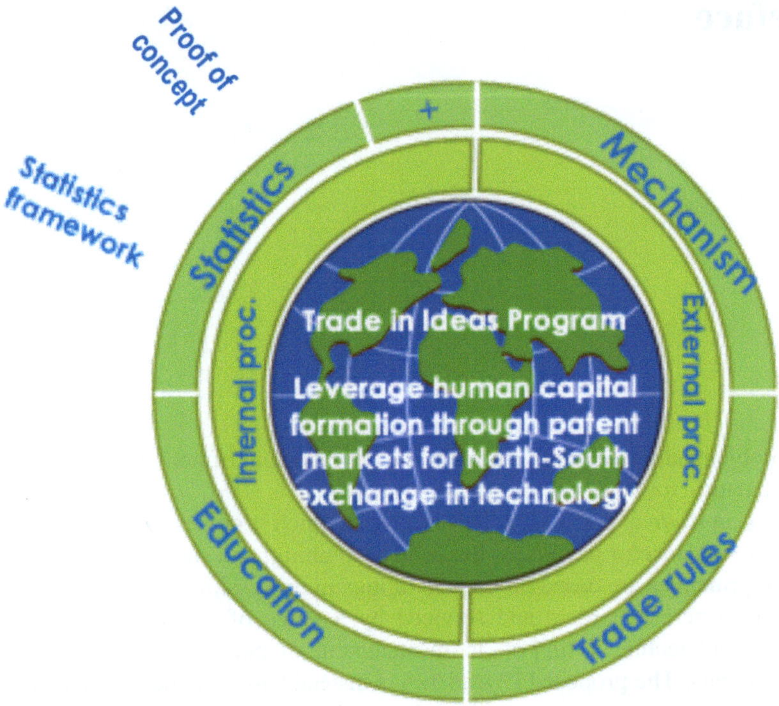

Fig. 1 Four themes covering a number of projects and structured by two processes, one external and one internal to the country, coordinated by one program. This book covers the "Statistics" theme and projects

This initiative for an improved statistics on trade in ideas (IP) is an integral part of the Trade in Ideas Program and an outcome of the initial findings of the pilot study of seven countries 2017–2018 (Fig. 1).

For further information on the Program, see www.tradeinideas.com

George Mason University
Arlington, VA, USA

Eskil Ullberg

Acknowledgments

This book is based on a report entitled "Conceptual Framework to Measure Trade In Ideas for Patent Licensing and Transfers," 2019, and a follow-on proof of concept study also in 2019.

The work on the foundational report was carried out by a core team of individual experts and highly experienced professionals in trade, trade statistics, accounting standards, and practices from a range of organizations. The work was carried out through a number of workshops and team discussions.

The theory outlined in the book builds on the author's previous work on markets in patents.

The book expands on this work, especially on the theory of value, and with more examples to better place it as a contribution to IP statistics – a measuring framework for standards and trade in ideas – with the purpose to better understand the economic system of the twenty-first century.

A special thank you to the following experts for their excellent contributions during the workshops and discussions, especially to the statistics chapter (Chap. 3) and financial accounting chapter (Chap. 4) :

Andreas Maurer, Chief, International Trade Statistics Section, Economic Research and Statistics Division WTO
Joscelyn Magdeleine, Trade in Services Statistics Section, Economic Research and Statistics Division WTO
Michael Wolf, Senior Advisor, National Accounts Department, Statistics Sweden
Christian Surtin, Foreign Trade and Balance of Payment, Statistics Sweden
Joacim Bergman, Foreign Trade and Balance of Payment, Statistics Sweden
Dr. Claudia Tapia, Director IPR Policy & Legal Academic Research, IPR & Licensing at Ericsson
Henrik Carle, VP Business Management & Finance, IPR & Licensing at Ericsson
Ann Tarca, Board Member, of the International Accounting Standards Board, UK
Henry Rees, Technical Staff of the International Accounting Standards Board, UK

A great thank you and acknowledgment to the team from Instituto Nacional de Defensa de la Competencia y de la Protección de la Propiedad Intelectual, INDECOPI, who made possible the proof-of-concept study in Perú (Chap. 8) both from a conceptual and practical perspective.

The Expert team from INDECOPI included:

Ivo Gagliuffi, President of the Board
Mauricio Osorio, Assistant Manager, patent promotion (main focal point)
Elizabeth Davila, Directorate of new technologies
José Zúñiga Avila, Directorate of new technologies
C. Fernandez Polo, Specialist, economic studies
P. Fuentes Ramos, Specialist, economic studies

A special thank you to the Government of Sweden who funded the referenced pilot study, the statistics workshops, and direct costs for carrying out the proof-of-concept through a grant from **UD/HI**.

Contents

Part I A New Theory of Value and the Need for a New Measuring Framework for Standards and Trade in Ideas

1 Introduction . . . 3
1.1 The Economic System . . . 5
1.2 The Integration of Science and Technology . . . 7
1.3 A Culture of Creativity . . . 8
1.4 Trade in Ideas – The World Market in Patents and IP . . . 9
1.4.1 The Institutional Environment . . . 9
1.4.2 The Economic Environment and Business Strategy . . . 11
1.4.3 The New Economic System . . . 12
1.4.4 Summary of Trade in Ideas and What Data Is Needed . . . 14
1.5 Measuring the Contribution of Trade in Ideas, an Overview . . . 16
1.6 The Theory of Value Summary . . . 18
1.7 The Need for a New Framework for Standards and Trade in Ideas . . . 19

2 Theory of Value . . . 25
2.1 Introduction to the Dynamic Economic System of Trade in Ideas . . . 27
2.1.1 Upgrading Trade Theory to Patents . . . 27
2.1.2 The Key Economic Question Is Sociological . . . 27
2.1.3 The Management of Risk and Uncertainty and How to Measure It . . . 28
2.1.4 Theory of Value of a Dynamic Economic System with Uncertain Assets . . . 30
2.1.5 An Empirical Model of Value Based on the Strategies Firms Use to Solve the Uncertainty Problem . . . 35
2.2 The Strategies Outlined . . . 35
2.2.1 The Strategies Which Resolve Uncertainty . . . 35
2.2.2 Reward Structure by Strategy . . . 37

2.2.3 The User's Values and Two-Tiered Prices 38
2.2.4 A Socially-Strategic Equilibrium 39
2.2.5 Non-satiation Postulate for Uncertain Values 41
2.2.6 Investment Strategy and Maximizing Value 42
2.2.7 Industrial Organization 43
2.2.8 Concentration of Capital 43
2.2.9 Integration of Science and Technology 44
2.3 The Institutions 44
2.3.1 Institutional Performance and Strategy 45
2.3.2 Price Equals Value 45
2.4 List of Variables in the Empirical Model and Their Meaning 46
2.4.1 Economic Environment 46
2.4.2 Institutional Environment 48
2.4.3 Data to Be Collected 50
2.5 Measures of Outcome and Outcome Behavior 50

Part II Today's Statistical and Financial Accounting Systems and the Additional Data Needed

3 Today's Balance of Payment Statistics and Data Gap 53
3.1 The Statistical Frameworks and Data Available Today 53
3.2 National Statistics: The Treatment of Embedded IP Today 54
3.2.1 Current Focus on is on Embedded IP 54
3.2.2 Globalization Requires Broader Institutional Scope 55
3.3 Concepts Pertaining to IP-Related International Transactions and Current Data Availability 55
3.3.1 The Economic Statistics Related on IP Today 55
3.3.2 The Classification of IP-Related Transactions Today 57
3.4 Measurement Challenges in GVCs 61
3.4.1 Recording in Practice 61
3.4.2 Lack of Practical Recommendations for Compilation 62
3.4.3 Residual Ownership and Standardized Methods 63
3.4.4 Special Purpose Entities and Proposals 63
3.5 Characteristics of IP Traders as "Merchanting of Services" 64
3.5.1 Network Cooperation and e-Commerce 65
3.6 Conclusions 65
3.7 The Gap in Data to Measure a More Dynamic Economic System 66
3.7.1 The Data Gap in Statistics 66
3.7.2 A Change in Structure Needed 67

4 Today's Financial Accounting Standards and Data Gap 71
4.1 A New Challenge for Management 72
4.2 The Accounting Frameworks and Data Available Today 73
4.2.1 Income Statement: The Principles for Recognition of Revenues 74

4.2.2 Assets and Liabilities: The Criteria for Recognition and Measuring the Intangible Assets . . . 80
4.2.3 Examples of Accounting Practices Today . . . 87
4.2.4 The Need for a New Accounting Definition of "Product" . . 91
4.2.5 The Need for New Accounting Tools . . . 91
4.2.6 Patent (IP) Assets and Liabilities on the Balance Sheet . . . 91
4.2.7 Summary of Recognition of IP Assets . . . 92
4.2.8 Summary of Recognition of Revenues from IP and of IP Assets . . . 92
4.3 The Data Gap Compared to the Proposed Theory of Value . . . 94
4.3.1 The Provenance of Revenues – New Definition of Revenue Types . . . 94
4.3.2 A New Structure to Measure the Return on IP Assets . . . 95

Part III The Statistical Framework for Trade in Ideas

5 Overview of the Framework . . . 101
5.1 The Economic Environment . . . 103
5.1.1 IP Revenues and their IP Asset Provenance . . . 103
5.1.2 IP Assets and Liabilities . . . 104
5.1.3 Strategies . . . 105
5.2 Institutions and Contracts . . . 110
5.2.1 Institutional Rules of Trade . . . 110
5.2.2 Contracts . . . 111
5.2.3 Regional and World Prices . . . 111
5.2.4 Payments . . . 112
5.3 Dynamic Economic Value Through Trade in Ideas . . . 112
5.3.1 Value from Trade in Ideas (Outcome) . . . 113
5.3.2 Measures and Index . . . 113
5.4 Data to Collect on the Economic System . . . 113
5.5 Scope of Data for Statistics . . . 114
5.5.1 The Statistical Unit and Data on the Economic System . . . 114
5.6 Standards for Data Collection . . . 120

6 Description of the Collection and Compilation of Data for Each Element . . . 127
6.1 Changes in Principles and in Data Collection Structure . . . 127
6.1.1 Principles . . . 127
6.1.2 Structural Changes . . . 129
6.2 New Data to Be Collected . . . 131
6.2.1 Price Data Collected from Contracts . . . 131
6.2.2 Strategy . . . 132
6.2.3 Institutions . . . 132
6.2.4 Summary Table of Data Collection . . . 132
6.2.5 Statistical and Financial Accounts . . . 132

6.3 Summary of Proposed New Measures . . . 143
6.4 A Trade in Ideas Index . . . 145

Part IV A Proof of Concept

7 A Proof-of-Concept Study in Perú . . . 151
7.1 A Developing Country Study in Perú . . . 151
7.2 The Design in Cooperation with INDECOPI . . . 152
7.3 Selection of Inventor Firms . . . 152
7.4 The Survey Structure . . . 153
7.4.1 The Operation of the Economic System in a Developing Country . . . 153
7.4.2 The Survey . . . 153
7.5 The Results . . . 157
7.5.1 The Economic Environment . . . 157
7.5.2 The Institutional Environment . . . 162
7.5.3 Additional Data Needed for Statistical Purposes . . . 163
7.5.4 The Outcome Measures . . . 163
7.6 Accounting Principles and Practices . . . 164
7.6.1 Accounting Principles and Standards . . . 164
7.6.2 Accounting Practices . . . 164
7.7 What Can We Learn from the POC? . . . 165

8 Can Data be Collected in a Developing Nation? . . . 167
8.1 An International Effort of Sovereign Nation States . . . 167
8.2 The Data Needs and Standardization Needs . . . 168
8.3 An International Discussion Needed . . . 169
8.4 The Primary Tools . . . 169

Part V Implementing the Framework

9 Future Steps Toward New Standards and Statistics . . . 173
9.1 The Way Forward – BPM7, IFRS15, IAS38, and Standardization Bodies . . . 173
9.2 Process for Adjusting the Standards . . . 174

Bibliography . . . 175

Author Index . . . 177

Subject Index . . . 179

List of Figures

Fig. 1.1 New measuring standards are needed for an economic system increasingly relying on IP assets, to create statistics on trade in ideas, especially patent transfers and licensing 4
Fig. 1.2 Logical categorization of transactions by modularity of productivity, human capital, and IP 6
Fig. 1.3 Internationally agreed property rights on ideas and communications establish the global markets in IP, leading to gains from specialization and economic development. Firms adopt strategies to minimize risk-uncertainty/return in every market 15
Fig. 1.4 The statistical system overview for measuring standards and trade in ideas 17
Fig. 1.5 Overview of the economic system and the measurement approach for financial accounting and statistics 20
Fig. 1.6 Restructured and upgraded balance of payment (currently at BPM6) including a new, separate, current account for "trade in ideas" covering all transactions in patents/IP and the embedded patent/IP value related to trade in goods and services (colored areas) 21
Fig. 1.7 Income statement and balance sheet to better reflect the idea-driven business 22
Fig. 1.8 Characterization of the idea-driven economic system 23
Fig. 2.1 The rules – whether norms or in law – negotiated by members of and enforced by the institutions embodying them should give incentives leading to a certain decision behavior (strategies) which in turn is socially desirable 31
Fig. 3.1 New focus on BoP to include more of the dynamics of the trade in ideas (IP) 69

Fig. 4.1 Empathic management of talented people needed to create competitive inventions for a market in ideas whose outcome, including specialization in certain technologies and financial Returns on IP Assets (RIPA), can then be captured in the financial accounts ... 73

Fig. 5.1 The statistical framework aims at capturing the dynamism of an idea-based economic system through expanded and differentiated standards for data collection, how data are compiled, and new measures for trade in ideas calculated ... 102
Fig. 5.2 The idea-based economic system used for the statistics framework ... 102
Fig. 5.3 Revenues are classified according to their provenance (assets used in their production) ... 104
Fig. 5.4 Access to IP as production rights are considered assets (for licensee and licensor) and liabilities (for licensor) ... 105
Fig. 5.5 Four strategies characterize firms' investment to manage uncertainty and trade behavior to manage risk in ideas: (1) staying clear, (2) strategic alignment, (3) marginal contracting, (4) systemic abuse ... 108
Fig. 5.6 Statistical framework components ... 115

Fig. 7.1 Inventors' economic environment ... 157
Fig. 7.2 Team sizes and legal entities ... 158
Fig. 7.3 Legal entities ... 159
Fig. 7.4 Inventor education levels ... 159
Fig. 7.5 Inventor teams age ... 160
Fig. 7.6 IP awareness ... 161
Fig. 7.7 Patent portfolio growth ... 161
Fig. 7.8 National and international trade in ideas differentiated by type of strategy: (1) Staying clear, (2) Strategic alignment, (3) Marginal contracting, and (4) Systemic abuse ... 162
Fig. 7.9 Assets exchanged for patents ... 163

List of Tables

Table 2.1 Strategies firms use to create trust in each other's actions 36
Table 2.2 Reward structures for the strategies discovered 38
Table 2.3 Investment needs to play the different strategies 43

Table 3.1 Treatment of intellectual property in BPM6 and MSITS2010 guidelines 58
Table 3.2 Reporting of BPM6/MSIT2020 intellectual property related items, 2015 59
Table 3.3 IP statistics 68

Table 4.1 Financial accounting to measure the performance of management of inventions 79
Table 4.2 The treatment of intangible assets 88
Table 4.3 Recognition of intangible assets 89
Table 4.4 The IP balance sheet 90
Table 4.5 Accounting data gap 96
Table 4.6 Accounting standard gap 98

Table 5.1 Part I: Standards for data collection on the economic system, 1. Agent, 2. Agent characteristics, 3. Products traded, 4. Strategy 121
Table 5.2 Part II: Standards for data collection on the economic system, 4. Strategy, 5. Institutions, 6. Contract 123
Table 5.3 Part III: Standards for measures on the economic system, 7. Outcome measures 125

Table 6.1 Principles on which data shall be collected 133
Table 6.2 Data to be collected on the economic environment and strategy 136
Table 6.3 Data to be collected on institutions and contracts, and summarized data 139
Table 6.4 Income statement: Current and proposed accounts for standards for revenues and cost 141

Table 6.5 Balance sheet: Current and proposed accounts for standards for assets and liabilities ... 144
Table 6.6 Measures for institutional and management analysis and ranking ... 146

Table 7.1 Summary of questions on revenues from patents/IP that would go into the IS and BS of inventor firms ... 154
Table 7.2 Questions relating to cost that would go into the IS and BS of inventor firms ... 156

About the Author

Eskil Ullberg is an Adjunct Professor at George Mason University and the Head of the Trade in Ideas Program. His research interest is on markets in patents, and how they can leverage the human capital formation, especially for developing countries, through exchange in human ideas.

Teaching areas include International Economic Policy, International Economic Law, International Finance, and African Economic Development.

He is a pioneer in studies of markets in patents using experimental economics. The work has, in its applied form, attracted attention from a large number of developing nations, focusing on market efficiency, a statistics framework on trade in ideas, regional field experiments in economic cooperation areas to evaluate trade rules, and educate the next generation of managers and policy makers in patent management. Prior to his academic work, Eskil worked as a strategy consultant for 20 years for companies, government agencies, and international organizations focusing on strategy and the management of risk and uncertainty. His work has been published in academic journals and books and has been presented at international organizations including UN (ECOSOC) and WTO, focusing on maximizing the human potential.

Part I
A New Theory of Value and the Need for a New Measuring Framework for Standards and Trade in Ideas

Chapter 1
Introduction

The operating of the economic system is changing. Ideas, especially technical ideas, and technologies are becoming the headline issues for managers and international policymakers and universities. *Trade* in ideas – exchange of intellectual property (IP) rights between nations' inventors – is logically the first part of this economic system equation. Trade first takes place in ideas, then in the goods using the ideas, and then in the servicing of the goods. The relative importance in international trade is now shifting toward ideas, especially IP protected ideas.

The ideas, and here the focus is on *technical ideas turned into tradable rights*, i.e., *patents*, are developed in the meeting of the *demand* for new solutions and the *supply* of ideas to create economically viable solutions. (Today this equation is rapidly expanding to socially, economically, and environmentally viable solutions.)

Trading nations do therefore not only benefit from each other's production possibilities, capital, and knowledge, but from each other's patented technology (based on specialized knowledge). These tradable IP rights create a market in ideas, allowing the selection mechanism to expand from a personal and goods-and-services embedded one to an impersonal, international, and non-embedded-idea one at Internet speed. Such markets promises gains from specialization in technology further increasing the development of the world's stock of technology.

The policy goal is therefore changing toward the social and economic improvement of the *institutional performance* and *behavioral properties* of markets in patents (and other existing and future IP). Such a policy framework needs to be informed by statistics on trade in ideas, which is shaping the economic system of the twenty-first century. It needs to integrate the trade in ideas, goods, and services into a whole with social and economic meaning.

The statistics need an upgrade to better reflect the operating of the economic system: *Ideas → Goods → Services.*

This book *proposes* a new intellectual property (IP) statistics: *a measuring framework for standards and trade in ideas* (IP). It takes its starting point in the fact that trade is *first of all* in the ideas, which are then used in the development, design, production, servicing, recycling, and sometimes financing and even trading,

E. Ullberg, *Intellectual Property Statistics*, Contributions to Economics,
https://doi.org/10.1007/978-3-031-36386-3_1

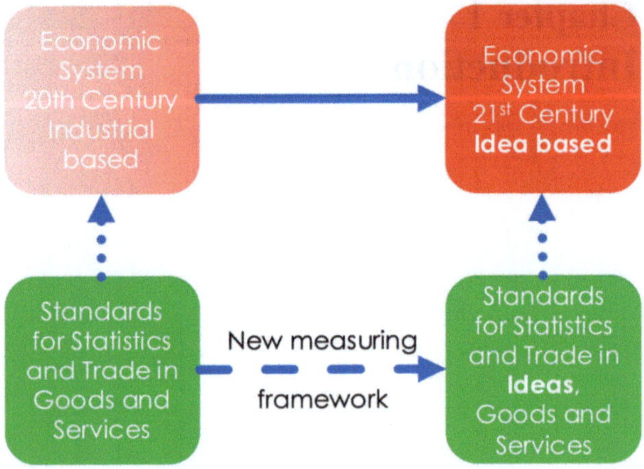

Fig. 1.1 New measuring standards are needed for an economic system increasingly relying on IP assets, to create statistics on trade in ideas, especially patent transfers and licensing

i.e., electronic commerce of new goods and services. The trade in these ideas is global since long and has created a highly dynamic economic system – where IP has taken an increasingly prominent role – and needs to be better understood to inform trade policy for the twenty-first century (see Fig. 1.1).

The approach thus differs from the current principle for statistics that is based on trading "*things in possession*," physical goods that are moved across borders, and then serviced by a range of services from guarantees, finance, and technical support to management of performance and risk-sharing arrangements. The proposed IP statistics is instead based on the principle of "*things in action*" being the key economic activity of the twenty-first century, focusing on patents (productivity and technology), i.e., the intellectual property *rights*. This has consequences for the economic understanding of trade value flows as the IP rights are much riskier and uncertain than other production assets.

The first section will therefore propose a novel theory of the value of trade in ideas attempting to better take into account these risks and uncertainties. It is followed by an analysis of the current standards and the data gap needed to be closed, compared to the proposed theory. The framework elements are then presented, followed by a detailed discussion, and capped with a proof-of-concept study from one developing country and an implementation strategy for *upgrading* current international statistical and financial accounting standards. The purpose of the book is thus also to discuss the implementation of the framework, would it be deemed important by users and producers of statistics.

The proposed statistics can be used by international standards bodies, by statistical compilers, to provide statistical information for regional and international negotiations on trade in ideas, i.e., polylateral and multilateral treaties and agreements, and, of course, the *inventors* – individuals, small teams, SME, large businesses, and universities/institutes – in their development of the international markets in patents (IP).

The statistics would therefore serve in the development of the international economic law corpus, institutionalizing the mechanisms making the market in patents (IP) more efficient. It should therefore be practically useful for both developed, least developed, and developing countries (already in the mid-2020s the developing countries may overtake the developed countries in human capital formation measured in peer-reviewed academic article publications, a critical source of new knowledge for state-of-the-art inventions). Although the book is focusing on patents, the same concepts can easily be expanded to other existing and to future IP as well.

1.1 The Economic System

The economic system can be seen as operating based on the principle of advancing productivity through technical solutions – *inventions* – to meet the ever-growing human needs.

These solutions are created by means of a highly dynamic *exchange* in human ideas. Some of the technical ideas are protected by intellectual property (IP) *rights*, mostly technology patents, creating a *market in patents* by transfers and licensing contracts. This impersonal *trade in ideas (IP)* allows inventors to specialize, and thereby increase the productivity in the *production* of the new technology as well.

These productivity gains in technology *production* from trade thus results in a faster growth of the stock of technology than without such exchange and specialization. This approach to the economic system can be seen as an "upgrade" of Adam Smith's trade theories, to patents (IP).

Typically, the exchange is not only for money but for *other* patents, *other* IP like data, trademarks, etc., equity, *other* assets, or market access (freedom to operate). Identical patents validated in several jurisdictions (countries) form a patent "family". Access to parts or whole patent families is how the trade in ideas (IP) is practically conducted. The productivity gains from specialization, in turn, increase the stock of technology and specialized knowledge (science) globally at a higher growth rate, benefitting all.

Thus, the patent and IP markets also provide a key mechanism to link science (research) and technology, further advancing the technology based on the advances in the stock of knowledge. These advances are triggered by the *demand* for scientific knowledge to resolve technical problems. The demand is also revealed through the published patent applications themselves. This "spill-over" is in fact an inherent *design* feature of the patent system, a social contract of sharing ideas in exchange for excluding and, more importantly, *tradable* (transferrable and licensable) property rights for industrial use, and a fee to the government. The patent system can therefore be seen as a key mechanism for *integrating science and technology as well as a trading system for ideas*.

The growth in the stock of knowledge provides the basis for a constant productivity growth as new technology development ultimately suffers marginal decline due to lack of new understanding of nature. The well of knowledge needs to be expanded to feed new inventions.

Since the patent system requires the inventor to teach the world about their invention by disclosing it *publicly* in a way that "a person skilled in the art" can "reduce it to practice," "exchange" in ideas can take place outside the patent *market* through theft severely infringing the patent holders' rights at Internet speed. A policy of forced technology transfer – technology handover in exchange for market access – is another way "exchange" takes place and infringes the *voluntary* aspect required for markets to work effectively. Hostile takeovers, bankruptcy auctions, and industrial and cyber espionage are other ways "exchange" takes place on current and future patents. This infringement aspect is present for all IP.

Despite these social, economic, and political issues robbing the inventors of their "honor", patents (IP) are increasingly traded under market conditions where both parties mutually agree on the terms of a contracted exchange. Because the disclosure requirement also informs inventors about the *state-of-the-art* (which was its explicit intention) and the possible intentions of future technology development by others, it makes it possible to fill "technology gaps," find specialized partner inventors, or "signal" intent for cooperative or "stay out of my technology area" strategies. For a study on what strategies the world's most patent licensing active firms use, see Ullberg (2015a). To say the least, this exchange in ideas is fiercely competitive, selective, and complex, and requires global strategies backed by global policy for global social and economic success.

Whether the ideas are first traded to create a pool (a "bundle") of technologies to sell or license, or used directly by the inventor firms, ultimately the ideas must be used (embodied) to create new goods and services *innovations*, thereby realizing the desired productivity growth. This innovation process requires a selection process in its own rights for which *inventions* to further invest in. Many market tests are often needed to make the innovations embodying the inventions economically successful. Often, this is "discovered" through a trial-and-error market-testing process, based on deep-seated beliefs about consumer behavior preferences, or in a serendipity fashion. **In the end, a good or a service can be seen as a well-selected *bundle of patents*.**

In trading these IP rights, each inventor benefiting from the specialized knowledge of the other, the productivity of technology itself is thus increased. **This highly dynamic exchange is at the *heart* of the economic system today.**

The economic system can then be said to operate in a market economy for ideas as well as goods and services. This creates *a logical categorization of transactions as: trade in ideas, trade in goods and trade in services.* Trade increases gains from specialization in ideas, goods, and services (see Fig. 1.2).

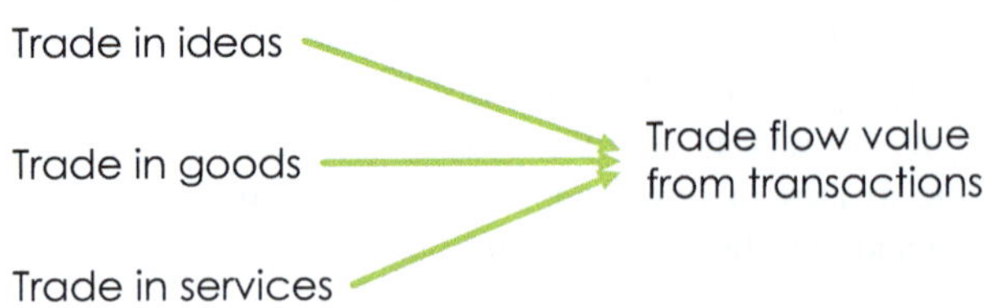

Fig. 1.2 Logical categorization of transactions by modularity of productivity, human capital, and IP

As an example, this market in patents (and other IP) has been central to the development of many new technologies today such as telecom, artificial intelligence (AI), and other ICT, energy, and food/clean water technologies.

However, the international financial accounting standards and public statistics such as the balance of payments (BPM6) still reflect the "industrial" paradigm of the early twentieth century based on capital (physical assets) and manual labor where manufacturing and, in the last half century also services, have been the main focuses of attention. Current standards have only partly, and incompletely, been updated in an *ad hoc* fashion, to meet some aspects of the new *transformative* business models and the global operation of the economic system of the twenty-first century. In reality, the economic system is today *dominated* by human capital formation of which some is turned into IP. Intangible assets (nonaccounted assets or nonphysical assets) represent up to 85–90% of firms' market valuations. They are included in at least 40% of the traded volume of goods, even higher in trade in services and "100%" in the trade in ideas (IP).

Without an upgrade and transformation of the international accounting systems, balance of payments and other statistics around the heart of the changing economic system, inventors, business managers, investors, and policymakers may draw the wrong conclusions about the state of the economy and thereby propose the wrong economic policies.

This book proposes new statistics and a new *measuring framework for standards and trade in ideas*. It is an *addition* to and a *restructuring* of existing accounting principles, statistics frameworks, and position statements with the purpose to better reflect the role and value of patents and other IP in international trade (see Fig. 1.1).

The framework will separate the value flows of trade in ideas (IP), trade in goods, and trade in services, creating the necessary statistic to better reflect the functioning of the economic system of the twenty-first century.

It is a clarification of the trade flows in the idea-based economic system by the *nature* of the economic development (rights based), which meet economic and social needs with an increased attention on preserving nature for the generations to come.

1.2 The Integration of Science and Technology

The markets signal to the research community about which problems are of interest for economic development. The *patent markets* therefore contribute directly to the *integration* of science and technology (Ullberg, 2012). Inventors seeking to develop new technology in response to market demands typically turn to research universities and institutes for supply of new knowledge. Patents (IP) allow these transactions to take place in an open market way. It was the integration of science and technology that created the "second economic revolution" ushering in the industrial era (North, 1981, p. 17).

By an integration through the patent (IP) system and patent markets, a *constant rate* of technology development (productivity growth) can be achieved as the stock of knowledge (science) is constantly expanded by solving the "right" kind of problems, compensating for the eventual decline in *marginal returns* on technology development if based on a *constant* (or declining) stock of knowledge. The patent markets thus act both as an important *selection mechanism* on what technology to further invest in and *what* scientific investigations to pursue. This function of the patent market is explored in Ullberg (2009, 2012, 2016) and from a least developed and developing country perspective in *The Trade in Ideas Program.*[1]

To measure the efficacy of this integration, we need to calculate *the transformation rate* of human capital formation (HFC; stock of knowledge) to patents (stock of protected technology). The framework therefore also needs to *separate* out the contribution of the IP assets on which transfers and licensing are based.

This means that a data collection on both the patents, their technology areas (e.g., by the International Patent Classification, IPC, code), and the stock of knowledge (e.g., by the OECD classification code) is needed.

An estimate of these returns was done during a pilot study of seven developing nations comparing them with 192 nations (Ullberg, 2018).

It was found that the world economic system operates using two "systems": one with IP rights and one without IP rights (in practice de facto that is, not *de jure*). The transformation rate *increased* with stronger HCF among IP using nations. The rates varied from 1/10 to 1/1 patent applications (peer-reviewed) per academic article, the measure used for HCF. At the higher end of HCF there appeared to be a "marginal decline," indicating *insufficient investment* in basic or in applied research, as indicated above, or *inefficiencies* in the transformation mechanism.

1.3 A Culture of Creativity

"Entrepreneurship is a rare flower."[2] With *inventors* it is similar. Some are one-off inventors and some serial inventors. These serial inventors are the "gold nuggets" in the gold field. They continue to invent, idea after idea, or they are inventor firms that continue to invent over many generations, having created a "culture" of inventions. Investments in R&D, new ideas, patenting (IP), and IP trade need to be institutionalized to make this integration of science and technology most productive. The ability to select the "serial inventors" to hire or fund and to create a *culture of creativity* then becomes a most important asset for generating the ideas and IP later to be traded. This is a very challenging management and policy problem.

[1] See www.tradeinideas.com for presentations, discussions, and updated information about the program.

[2] Quote attributed to V. Smith.

There are many stories about unhappy inventors running off and starting a new venture when the firm they work for are "satisfied" with the first idea. A great example is Seymore Cray, the designer of the CDC 6400 "supercomputer" in 1964 after having turned computing from tubes to semiconductors. In the pursuit of even faster computer, he quitted and started Cray Research in 1972, in part working for defense contractors, with "acres" of Cray-1 for cold-war needs. Then in 1995, after the end of the cold-war with less strategic needs and the arrival of the PC, the company went bankrupt, but he started SRC Computers in 1996. His last work was on Gallium-Arsenide, which he thought would be the future semiconductor materials for even *faster* computers. These individuals or small teams have huge impact on whole industries, i.g., information technology, bio-tech, and their technologies on peace and prosperity. It may therefore be important to track these serial inventor's patents as an *indicator* of a "culture of creativity" in firms, regional clusters, or nations.

1.4 Trade in Ideas – The World Market in Patents and IP

Trade in ideas, as any other trade, takes place in an *economic environment* of agents who have different agent characteristics (IP, resources, knowledge, and strategic choices) and contracts (transfers and licensing), and an *institutional environment* of idea, and exchange and communication (property) rights, which governs the operation of the market. Together these environments form the *economic system* (cmp. Smith (1982)). The description of the market in patents follows this division in economic and institutional environments to create a schematic overview of the new idea-based economic system useful for measuring the economic and social contributions of trade in ideas by means of international accounting standards and public statistics.

1.4.1 The Institutional Environment

The transformation mechanism of human ideas into economics through markets in patents (IP) is mainly an *institutional policy issue.*

Two Types of Property Rights

The institution defines the rules of property rights on *ideas* (patents, other IP) and rights on *communications and exchange* (messages buyer, sellers, traders are allowed – by rules or norms – to send during negotiations, rules governing these communication right, and rights to claim ideas (IP) or payments according to contract agreements) under which they are *transformed* (such as patenting of ideas) or *exchanged* (voluntary contracts, government privileges to strategic technology, nonvoluntary forced tech transfer, etc.).

The institution therefore defines the right to speak/not to speak during an *exchange* (in negotiations, in courts, etc.), the right to demand payment (for the use of rights, for example university privileges, right to contract with national or foreign private or public owners, etc.), or transfer of rights (between firms, from university to investor, etc. given the agreed terms) and the right to exclusive use (private ownership of the patented idea).

The *transformation* of technology to patents (IP) also has rules of private "ownership." These are thus codified in patent (IP) laws. The intellectual framing of what defines the IP rights used here is mainly the "minimum standards" related to the TRIPS agreement.

To evaluate the *institutional performance and behavioral properties* of the markets in patents (IP), key characteristics of these institutions need to be measured. Statistical measures therefore need to characterize these *patents* (IP) and the *international trade rules of exchange*.

Property Rights on Ideas

The rules on patent rights can range from "automatic injunctions" in case of infringements as in Germany, to virtually "no injunctions" as in the United States. This rule, changed by the Supreme Court in the famous "eBay case" from 2006, "has gutted the system" in the United States, to quote an Hon. judge, as there is now little incentive to "settle" between the parties infringing. The German rule, on the contrary, brings the parties to the negotiating table immediately as there is a strong incentive to clear the market in patents through settlement, usually by a license. (The Chinese system is based on the European system (EPC) and the German court system.) These are thus examples of property rights on the enforcement of *ideas*. The criteria on which inventions a patent should be granted, number of claims, disclosure, etc. are examples of the patenting process rights. The patent system consist of the patent granting office and an (independent) patent court. Some systems also have an administrative opposition procedure (an in-house post-grant "court"). All these rights construe the property rights on the ideas.

Property Rights or Rules on Trade

In Chile the universities typically do not get paid by businesses for IP because this is not the "culture" to do that according to a communication at the pilot-study workshop mentioned above. Universities are instead seen as mainly teaching institutions. However, today universities are producing valuable patented technology, often in cooperation with international universities, thus becoming *economic* actors. This is thus an institutional policy on research and education.

There are also limitations in some free trade agreements (FTA) on transfers and licensing of rights and from government-funded projects. These are examples of property rights on *communications and exchange of ideas*.

The importance of these seemingly small changes in rules with dramatic effect on *incentives* to trade requires measurement and characterization of the *trade rules* for ideas and the *patent rights* on ideas to understand the policy implications on trade in ideas.

Characterization of Property Rights and Trade Rules

The characterization of the *property rights on ideas* can be done based on patent laws. There are already many indices of the de facto and *de jure* operating of the patent systems worldwide, which are useful for the purpose of statistics.

The characterization of the *trade rules* for ideas can best be done by investigating general trade agreements,[3] regional free trade agreements, other international agreements and conventions, and categorizing the trade rules by the *constraints* that are placed on the trading parties' "communications and exchange" rights. Other sources are also important to codify social norms. These "constraints" structure the economic, social, and political relations (cmp. North (1991)). Statistics can then be created describing the institutional performance and behavioral properties of the *actual* markets, given this list of *constraints* (norms and trade rules).

1.4.2 The Economic Environment and Business Strategy

The economic environment covers the agents and agent characteristics, and the contracts on the transfer and licensing of patents (IP). It also covers the *business strategies* patenting and licensing active firms use. This is thus a new element to the theory: the allocation of resources for invention is dependent on the choice of business *strategy* of agents which, in turn, depends on the institutional environment just described. This point will be developed further.

Types of Agents

The type of *agents* ranges from individuals, small teams, SMEs, large firms, MNCs, and universities. They are characterized by having access to very different resources, including knowledge, technology, patent portfolios, geographic presence, technology focus, sales, age, nationality, research programs, business–university collaborations, and so on. These financial and other measures of patent (IP) holding agents are critical in measuring returns on IP assets. Much data exists in "silos" but data that make the connection between agents' financial statements, human capital formation, and patents is limited. This remains a practical challenge in characterizing the economic environment patent active agents operate in. In the referenced proof-of-concept study, a first connection was made using private database, indicating that it can be done. The purpose of statistics is to make (at least) the summary data public.

Types of Contracts

Contracts can also be characterized in a range of types (transfers, licensing, cross-licensing, etc.). Contracts often take months to negotiate but end up in quite similar fashion in the end.

[3]Discussions with patent law experts has not resulted in any mentioning of such agreement, so the conclusion is that no such agreements exist. There is a "lacuna" in the international trade law, as one diplomatic counselor to WTO said.

They are, as all contracts, "incomplete" and therefore the decision rights on residual rights to assets granted the contracting agents are important in the characterization (cmp. Hart (1988)). Statistics can therefore be developed by the type of contracts and decision rights, indicating the *direction* of the trade in ideas. This direction – an important element in trade theory and analysis – indicate the technology area *specialization* of a nations' inventor agents. The gains from trade-in ideas due to specialization in technology can then be summarized.

Types of Strategies

The global markets in patents (IP) are also *highly risky and uncertain* (in addition to difference in agent resources and incompleteness of contracts) and requiring business *strategies* by inventors and inventor firms and *policies* by governments including international *economic* law, i.e., international treaties and agreements, to be addressed. These strategies and policies thus aim to reduce the risk and uncertainty to levels where the markets *can* be efficient in allocating patented technology, deliver gains from specialization in their production, and create a high level of integration of science and technology.

If the risk and uncertainty is too high in trading, a separation of invention and innovations in different (competing) firms is reduced and the gains from specialization are reduced. The integration of invention and innovation then takes place in a single hierarchy (firm).

However, markets provide the more efficient allocation of resources for invention as a *competitive* selection of which *all* ideas to further invest in then takes place.

The *characterization* of agents, strategies, and contracts, including those between science and technology agents where IP is involved, using *standard types*, provides statistics for key elements of the economy and a way to measure the overall economic system performance.

Taxation Policy

The integration of science and technology (to provide new knowledge) is related to *taxation policy* including *education/research funding policy* (tax deduction of private research, private university donations, rights to publicly and privately funded research) in addition to institutional policy.

Tax policy is a main driver of transfer pricing and tax planning. Tax and education funding policy on IP is therefore important in characterizing the economic environment and needs its own statistics to "separate out" the transfer prices and tax optimization from the third-party transactions.

1.4.3 The New Economic System

The economic system of today is more dynamic than 50–100 years ago with international (global) cooperation in developing patented technology and learning, global value chains in goods and increasing digitalization creating Internet-based services. It is therefore not enough to measure the allocation efficiency in a *static* system of goods and services innovations based on the existing technology as an "external" input (externality). The institutional performance and *behavioral properties* of markets in patents (IP) need to be taken into the center of measurement standards and statistics.

Measuring a Dynamic Economic Efficiency

A dynamic economic system is a system where *learning* is part of the economic system. Both the outcome (the use of the patented technology) and input (creation of new patented technology) are dynamic processes. In these processes markets in patents (IP) are the selection mechanism of which (patented) ideas to further invest in.

What is needed to be measured is therefore the *system efficiency* with *dynamic* allocation of resources for *invention* (and innovation but that is a subject of global value chains). Efficiency should thus be calculated as a "selective output of how these inventions are used based on market tests" divided by a "selective input of which technology to invest further in."[4] One such desirable outcome is an intense cooperation between developed, least developed, and developing countries based on each other's own ideas and human capital formation.

The Statistical Unit and Trade Flows

Since inventions are typically made by individuals or small teams and there are now 100s of millions of highly educated people in the developing world, it is essential that the measurements can take into account ideas down to small or individual units of statistics. The "statistical unit" would therefore ideally be these individuals or small team corporations. The economic performance – output divided by input – is thus directly influenced by the institutional arrangements for these markets.

The need is then to capture these input and output dynamics of the trade in ideas process to inform policy. This includes data on trade patterns based on *trade flows* (export–import). In turn, measuring these flows requires a *theory of trade value* for trade in ideas.

These results can then lead to a better integration of the patent system (IP) into the economic system by (1) informing trade policy (discussion) and (2) transforming the WTO, general trade rules to give incentives for developed and developing countries alike to participate.

Trade in Ideas as an Own Element

The trade in ideas today is essentially "unregulated." Mainly the patenting process is "regulated" through TRIPS, EPC, and other international, regional, and national conventions. New trade rules are *needed* to better integrate the patent system in the global trade system. This integration needs to be a level playing field for North–South in order to be relevant with respect to the inventors developing the technology. New international economic agreements are therefore needed in the areas of trade, investment, contracts, and possibly other, streamlining the existing treaties. Twenty

---

[4]See Ullberg, "Trade in ideas as a new development policy", Trade Dialogues Lecture Series at WTO, June 30, 2017.

years of fighting over TRIPS among WTO members indicates that something is clearly not working with North–South relations in relation to IP. The upside is perhaps somewhat "muted" by the current de facto regimes. Statistics based on the performance of the economic system including learning appear therefore necessary to inform such policies of trade in ideas.

The previous discussion on the economic system and necessity to include trade in ideas as an own, separate, element is now summarized in a schematic overview. This overview is the basis for the proposed statistical framework.

1.4.4 Summary of Trade in Ideas and What Data Is Needed

This section summarizes the discussion in the introduction.

In trade in ideas the economic environment consists of *inventors* of different types (individuals, small teams, firms, universities, etc.), transfer and licensing *contracts*, and assets to trade (money, patents, other IP, equity, etc.). The *strategies* which firms use – the specific allocation (choices) of resources for invention – is thus part of this environment. The *institutional* arrangements governing the exchange, whether based on culture and norms or formalized in a written international agreement, are *essential* to produce efficient markets. There are arrangements both for the *creation* of the rights – the patent granting and enforcing procedures – and the *trade* in the rights, the exchange of rights by mutual consent.

Internationally agreed property rights on ideas and communications establish the global markets in IP, leading to gains from specialization and economic development. Firms adopt strategies to minimize risk-uncertainty/return in every market.

To measure the effectiveness of the institutional policy (and taxation policy where relevant), we need to measure the outcomes *given* these actual communication and exchange rights (see Fig. 1.3).

A "Creative Culture" creates incentives to invent new technology by protecting the inventors' ideas through tradable IP rights. This policy "honoring the inventor" exists since the first known patent system in Venice from 1474. "Sovereign nation states" then create international trade agreements with incentives for *cooperative strategies* between inventors and innovators as described earlier (economic environment section). An economically efficient trade in ideas then depends on the strategies firms choose, given the risk and uncertainty resulting from the patent system quality in practice *and* trade rules (institutional environment section).

The framework must then *separate* the outcome from agent types, strategies, and contracts from property rights on ideas and property rights on communications and exchange, i.e., the economic and institutional environments. It also needs to include a formalized description of the agents. These are tied to the *legal entity* used (individuals, firms, universities, etc.) and the characterization of their strategies, which determine the allocation of resources for invention (cooperative,

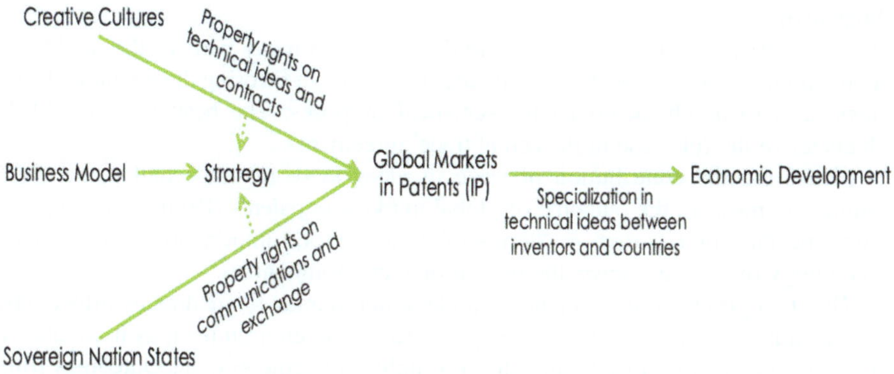

Fig. 1.3 Internationally agreed property rights on ideas and communications establish the global markets in IP, leading to gains from specialization and economic development. Firms adopt strategies to minimize risk-uncertainty/return in every market

noncooperative, or abusive strategies developed below). This allows to relate desirable outcomes, such as allocation of IP rights to their highest value use, and specialization of invented and patented technologies between inventors and countries to the specific institutional *rules*. This framework can then create relevant statistics to inform an efficient institutional policy.

For example, a university which has received a government grant for research may not be allowed to sell their technology patents abroad or has first to compensate the funders in order to do this (this naturally goes for private institutes or firm funders as well). This is important as most inventions take place in universities in the developing nations. Another example is that individual inventors may not be able to defend their rights in courts, eliminating the possibility to "secure" their rights often required by a licensor or investor. The cost of funding may be too high.

The global market operates on a complex web of internationally agreed property rights on IP (the TRIPS agreement of WTO in 1994, the European Patent Convention, EPC, an international agreement for European states of 1978, the Paris Convention of 1883, etc.) *and* on some, but much weaker, communications and exchange rights (trade rules) embodied in free trade agreements and other international economic law texts. These "international trade rules" can therefore be seen as mostly unregulated when it comes to patents. The actual situation thus relies more on *informal* norms and culture. As an example, the role of universities has changed, and they have increasingly become economic actors where firms pick up educated people as well as their (patented) ideas (Ullberg, 2015b, 2019) but trade rules are lagging.

In general, "honoring the inventor," the principle of the patent system since its creation, is weak *in practice* even if some treaties have that in the legal text. Trade rules that better integrate the patent system into the trade system are needed at the international level, by necessity transforming the WTO as well as regional trade institutions (Phelps & Ullberg, 2018) to be efficient for an idea-based economic system.

Measures

A set of new norms and trade rules are developing in many parts of the world, but more efficient ones are *needed* to structure the exchange favoring the inventor. These rules need to be characterized for statistical purposes. The best way may be to characterize the rules use in the actual trade agreements.

Measuring the contribution of trade in ideas to world trade and world GDP requires to measure the outcome of global markets in patents (IP) in terms of gains from specialization – such as "terms of trade," export/import prices of patented technology over time – *given* the institutional environment.

The strategies firms use whether to trade or not to trade IP are directly affected by this institutional environment, not only the economic environment, as it affects the risk and uncertainty in trade, making it a highly dynamic system. Outcomes from types of strategies can then be compared given the types of rules.

To measure these trade flows, one needs to gather private data from the inventors and public data on patent and human capital formation. This is discussed next.

1.5 Measuring the Contribution of Trade in Ideas, an Overview

The importance from a measurement and statistical perspective is then to provide these *dynamic* measures of the learning, technology, and patenting and its integration (allocation) through markets. Cooperative incentives are needed because inventions are typically local and made by individuals or small teams of inventors.

The negotiations in the global market in patent (IP) result in contracts between inventors, and between inventors and innovators. These are incorporated in legal entities – typically individual/team based – SMEs or large firms, universities, and so on. These contracts are the primary data source for *microeconomic* data on the firm's economic activity such as patent transfers and licensing and prices. This data is thus a matter of *international financial accounting standards*, a subject that will be discussed in a separate chapter. *Macroeconomic* data on the firm's patent holdings and specialized technology areas (stock of technology) and human capital formation (stock of knowledge) at a country level is the other main source of data.

The legal entity is chosen as the "statistical unit" about which data is gathered in the statistical framework. This choice allows a distinction to be made between trade rules given any legal constraints to trade.

The firms' technology area investments, types of transaction, and the type of legal entity may be used to identify the types of strategies they use. The strategies are characterized as "basic" or "standard" types of cooperative and noncooperative strategies (see separate section for proposed characterization).

The institutional rules for international trade are defined by regional free trade agreements, treaties, general agreements, or international law standards, which can be collected by using some indices on the institutional level. Special studies on and

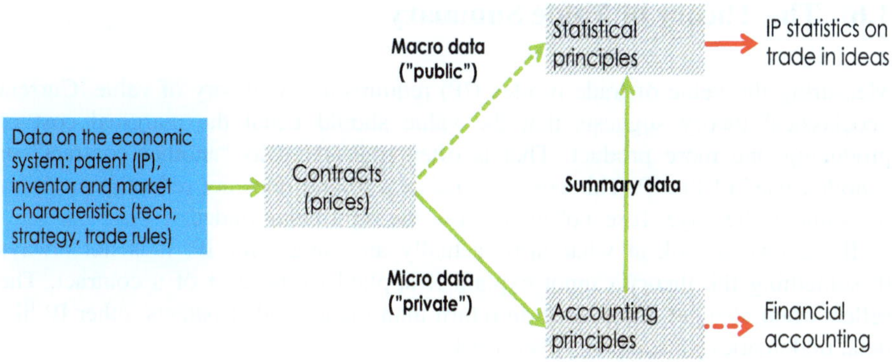

Fig. 1.4 The statistical system overview for measuring standards and trade in ideas

data collection by customs and other sources can also be used on what treaties are actually used. The importance of the institutional arrangements will likely require at least the summary of existing legal trade documents and standards or special studies with this *trade economic* angle.

The "general types" of treaties can be made as patents/IP (traded), trade secrets (not openly traded), carve-outs, and free flow of ideas (not protected or post-protection). Forced transfer, industrial espionage, and theft are other "types" as are tax optimization and grant (funding) motivated activities. These are part of the economic system.

Together the data can then be summarized to form the information flow for statistics (see Fig. 1.4).

The *contract* that specifies the exchange of assets and rights to assets in the trade in ideas is the basis of the financial accounting. This information is private.

For data from the financial accounts, i.e., *contract*-based transactions, statistical agents sample or send surveys, or sample or extract data from tax authorities on income statements and balance sheets. This data is the *anonymized and summarized* to form the trade patterns in terms of value flows between companies, countries, technology areas, and asset and rights exchanges.

Patents and Contracts as Production Assets

Importantly, the *patents* (IP) and also the *contracts on patents* (IP) are capitalized in the accounting and then depreciated and annually revalued (given applied research project status). *This allows patents rights to be put on the balance sheet as both assets and liabilities.* The patent rights – patent ownership, licenses, and license obligations – are therefore seen as "production assets" with a best-before-date, which are used or potentially used in new goods, services, or re-traded.

Anonymized Data

Another key principle is that the data is anonymous. To implement this principle, it is not enough to anonymize the data, as large players in small markets can still be identified. This is a major issue to practically resolve through standards.

1.6 The Theory of Value Summary

Measuring the value of trade in idea (IP) requires a new theory of value. Current neoclassical theory suggests that the value should equal the marginal cost of producing one more product. That is often interpreted as "another contract" or "another use" of the IP. This cost is close to zero and does not reflect in any way the value to the buyer (user) of the asset or the right being traded.

If we take a look at what firms actually are paying for IP, it is not always 0, something the theory cannot explain if applied to the cost of a contract. The sellers often also get other assets in return than money: other patents, other IP like data, trademark rights, equity, even work.

The proposed theory takes the starting point in the willingness to pay (WTP) by buyers/licensees and the willingness to accept (WTA) by sellers/licensors. The WTP is reflected in the value the firm can receive from *using* the technology in new goods and services, the value from *blocking* others to use the technology to compete, or the value from *(re-)trading* it together with other assets in a "technology portfolio" and the like. *The price paid is then equal to the highest of those values.*

The WTP and WTA is then dependent on the institutional environment given the incentives to trade. Institutions matter because rules matter, and rules matter because incentives matter.[5] If the incentives are to trade, i.e., to cooperate, then people will trade, and if they are not, i.e., to coordinate in a hierarchy (own firm), people will not trade. Thus, the price depends on the trade rules.

$$\text{Price}\ (\text{WPT} = \text{WTA} \mid \text{Choice of strategy}\ (d) \mid \text{Institutional arrangements}\ (r)) = \text{Value}$$

$d =$ type of strategy, $r =$ type of rule.

There are two rights granted by the patent system: the *right* of exclusive use and the *right* to transfer and license, creating the market in ideas, but to have a global market in IP, one also needs level playing field trade *rules*, which structures the communication and exchange in the IP down to the individual/small team. These rules are thus *outside* the IP laws, establishing the property rights in ideas. They are part of other and new international economic laws beyond GATT (goods), GATS (services), and TRIPS (IP related to trade in goods and services). To further the understanding of markets in patents, we need to further understand the impact of trade rules as well as the IP regime on outcomes. The actual operation of the trade system for ideas today is reserved to "the powerful few."

It Starts with the Contract

From a measurement perspective, the price is in the contract. Thus, through surveys of trading inventors and central bank transactions, one can assess the value.

[5] Quote from Vernon Smith, 2002 Recipient of the Sveriges Riksbank Prize in Economic Sciences in Memory of Alfred Nobel.

However, as cyber cash is increasingly used in transactions, and NFTs are issued in relation to patents, suggesting **that financial accounting is the *only* way to accurately measure the value of trade flows from ideas (IP)**. This is further emphasized because of the large "barter trade" in IP for other assets than money. The accounting principles proposed is therefore that all value measures start with the *contract*. This principle is *the same* principle used in the recent international financial accounting standard (IFRS15, recognition of revenues).

1.7 The Need for a New Framework for Standards and Trade in Ideas

The current frameworks for measuring trade flows focus almost entirely on goods and (increasingly) on services but only to a limited and incomplete extent on ideas, the *heart* of the increasingly idea-based economic system. The framework is summarized in figures.

The proposed framework takes into account the following:

1. The *separation* of trade in ideas from trade in goods and services to understand the impact on such trade on *the sequence of events* in *growing* of the stock of inventions/knowledge, productivity, and wealth
2. The *economic* environment of *specialized* technology producers, increasing the productivity of patenting technology
3. The *high uncertainty* in development and use of technology, requiring understanding the *strategies* inventors use in investing in IP to resolve the uncertainty to be able to trade
4. The *institutional* environment required to understand the highly dynamic economic system's *efficiency* in patenting technology and new learning
5. The *taxation* policy which creates "trade flows" based on tax optimization and policies on research which structure the location of patents.

There is thus the need for a new framework on measurements for standards and trade in ideas. This is done through micro data from contracts (financial accounting standards) and macro data from a range of patent, human capital, and other sources (statistics). A summary of a above is presented in Fig. 1.5.

Explanation of the elements are as follows:

I. Economic and institutional environments

   1. Patents and other IP
   2. Strategy – key issue
   3. Trade principles and rules
   4. Global market in IP

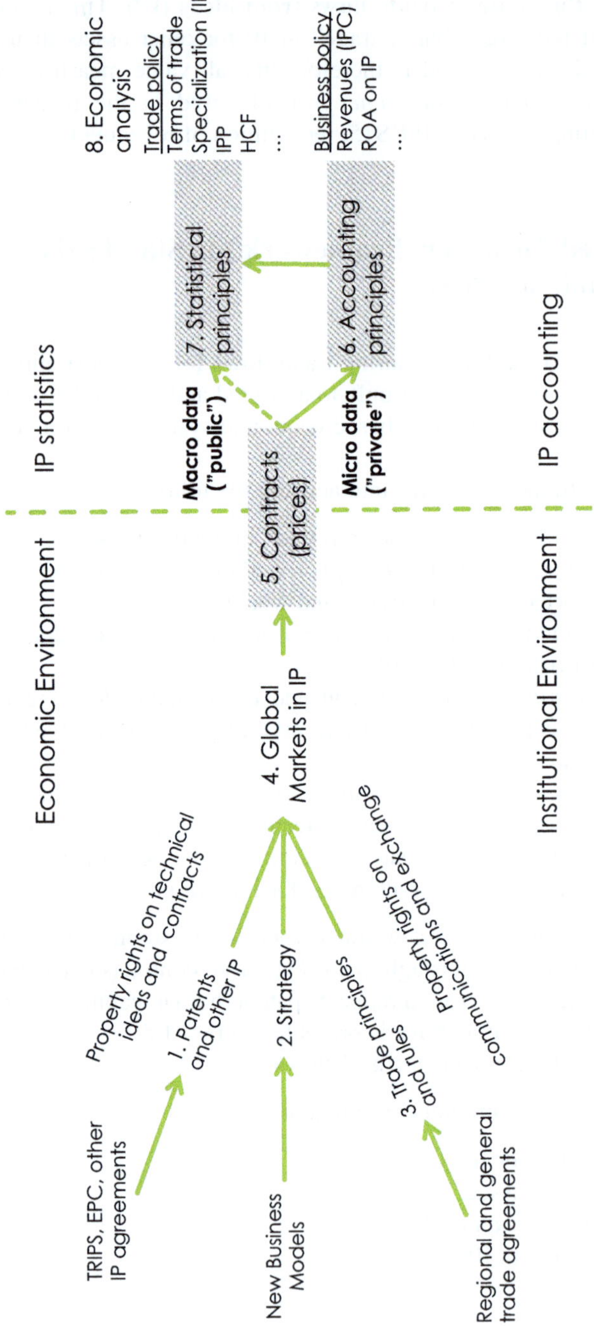

Fig. 1.5 Overview of the economic system and the measurement approach for financial accounting and statistics

II. IP statistics and IP accounting

5. Contracts – types of
6. Accounting principles
7. Statistical principles

III. Informing policy

8. Economic analysis

The stated goal is to inform economic policy on trade in ideas. The outlines of some new, key economic measures are:

(a) Trade in ideas current account (export–import | technology area)
(b) Terms of trade in ideas (export/import prices | technology area)
(c) Gains from specialization in inventions ($RIPA_t - RIPA_{t-1}$ | technology area)
(d) International Patent Position (ownership cmp. licensing of IP in use)
(e) RIPA (Return on IP Assets, ROA on IP)
(f) Technology specialization (International Patent Classification, IPC)
(g) Type of entity owning, using IP (industry structure; new IP entity)

The restructuring of the balance of payment and international financial accounting are outlined in a summary form in Fig. 1.6. Both patent/IP assets and liabilities

Proposal for restructuring of the balance of payments to better reflect trade in ideas	
Current account (income and savings)	**Capital account (change in assets, liabilities)**
Trade in ideas (things in action)	**Transfers of patents/IP (assets)**
Non-embedded patents/IP	**Cross-licensing of patents/IP (contract assets)**
Licensing	**Investment grants on patents/IP (capital transfer)**
Cross-licensing (net income)	Capital transfers and nonproduced nonfinancial assets
Bundles (portfolios, standards, etc)	
Abusive licensing (forced but legal)	**Financial Account (for assets, liabilities)**
Embedded patents/IP	**Direct investment in patents/IP (equity, debt)**
Embedded in goods	**Portfolio investment in patents/IP (shares, debt sec.)**
Embedded in services	***Financial derivatives linked to underlying patents/IP***
Trade in goods (things in possession)	Investments, reserves, and financial derivatives
Trade in services (productive activities)	
Primary income (savings)	**Statistical discrepancy (errors)**
Direct investment income on patents/IP	**Net investment - current - capital account balance in patents/IP**
Portfolio investment income from patents/IP	Net lending(borrowing) from financial - current - capital accounts
Other primary income	
Secondary income (grants)	
Grants on patents/IP	
Other secondary income	

Fig. 1.6 Restructured and upgraded balance of payment (currently at BPM6) including a new, separate, current account for "trade in ideas" covering all transactions in patents/IP and the embedded patent/IP value related to trade in goods and services (colored areas)

Proposal for restructuring, expanding the financial accounts to better reflect the idea business	
Income statement	**Balance Sheet**
Revenues	**Assets**
Licensing from nonembedded patents/IP *)	Non-current assets
Licensing from embedded patents/IP	...
Sale of patents/IP	Intangible Assets
Goods	Investment in/sale of own patents/IP assets (produce)
Software	Investment in/sale of group patents/IP assets (transfer)
Services	Investment in/sale of third party patents/IP assets (buy)
Total revenues	(Right-of-use assets extended to contracts on IP)
COGS	Contracts on patents/IP (inlicensing) (rights as assets)
R&D expenses	Financial assets
Patents/IP expenses (own, bundles, etc.)	Patent/IP backed securities
Other operating income and expenses	**Current Assets**
Operating income	...
Depreciation of patent/IP assets	Contract assets
EBIT	Receivables on patents/IP contract assets (outlicensing)
Interest payments	
Tax	**Equity and liabilities**
Net Income	**Equity**
	Non-current liabilities
	...
	Provisions, non-current
	Contracts on patents/IP liabilities (outlicensing)
	Current liabilities
	...
	Trade payables
	Payables on patent/IP contract liabilities (inlicensing)

Fig. 1.7 Income statement and balance sheet to better reflect the idea-driven business

are accounted for, including the (licensed) rights to the technology, enabling the calculation of the International Patent Position (IPP), an "International Investment Position" for who owns the patented technology and rights in use.

The income statement is further differentiated to better reflect the needs of management of inventions needed in an idea-driven business model. In particular, the patents/IP and contract on patents/IP are considered investable and depreciable production assets and liabilities (see Fig. 1.7).

The revenue streams from patents/IP and assets and liabilities are further differentiated to characterize the economic system from a cultural, business-market, and international economic (treaty) angle (see Fig. 1.8).

Economic system characterization	
The basis for an index on trade in ideas together with transactions data and balance of payments data	
Creative culture	**Business entity**
Cultural norms on honoring the inventor	Business strategy type
Identification of serial inventors/teams, companies	Type of legal entity developing,using the patent/IP
Education policy	Human capital formation, OECD education codes
The patent/IP items are further differentiated by	**Contracts and treaties**
Technology area - IPC classification	Type of contract
Field of use - product markets technology is used/licenced in	How "decision rights" are dealt with in the incomplete contracts
Geography - national markets the patent/IP are validated and licences in	International treaty obligations governing the transactions
Time - the length of patent life or licence	Taxation policy

Fig. 1.8 Characterization of the idea-driven economic system

This economic system characterization can be the basis to create a country and company *index* on trade in ideas.

The first version of the proposed framework for measuring and standards for trade in ideas is thereby complete.

Chapter 2
Theory of Value

This chapter will outline a theory of value for international trade flows in patents/IP and patent/IP licensing. What characterizes this new trade economic system is the *uncertainty* in the values of the patents/IP and the licenses traded.

Trade is part of *a highly dynamic economic system* of international trade in ideas (IP), trade in goods, and trade in services (embodying or not IP rights). Trade in goods and services is guided by international economic law (GATT and GATS) but is weak or nonexistent on trade in ideas where only "minimum standards" exist for the protection (not trade) of ideas (TRIPS). There is currently a lacuna in international economic law in the area of trade rules for ideas (IP).

The *institutional* arrangements are therefore of key importance for any theory of value as they help structure this trade, reducing risk and uncertainty.

The agent behavior provided by these institutions results in *strategies* to further reduce both risk and uncertainty. Without reduction of uncertainty there will be no trade, no specialization, and no gains from trade. The search capabilities reduce information risks. The human capital formation of firms reduces knowledge risks. The contracts, which often have a two-part tariff with a fixed fee and a royalty, reduce price risk. But is it primarily strategies and institutions that are necessary for the reduction of uncertainty?

The economic system description used in a theory therefore needs to take into account these elements of risk and uncertainty reduction in order to understand the value flows better. However, economic theory does not yet include uncertainty, only risk is included to some extent, and therefore an upgrade is proposed to better deal with this central theme to trade in ideas.

The theory of value will define the data needed to be collected to measure the value of cross-border flows in IP useful to inform business and government policy. The gap between the existing data collection and the new data and the new data collection mechanisms needed will then become the basis for creating the framework for standards and trade in ideas.

E. Ullberg, *Intellectual Property Statistics*, Contributions to Economics,
https://doi.org/10.1007/978-3-031-36386-3_2

The proposal brought forth is that (1) the genuine uncertainty that characterizes the economic value of patents/IP *first* needs to be reduced through the strategies the inventors use, and the international treaties agreed on, thus characterizing both strategies and institutions (treaties) as integral elements of the new theory of value; and then (2) exchange can take place when only risk is involved, in an economic environment using a standard economic approach.

The value can therefore be said to be a function of the *strategies* the firms use in a competitive economic environment, which in turn are a function of the *institutional* arrangements (rules, laws) used to reduce risk and uncertainty and make the markets more efficient. The characteristics of the economic environment – agents, agent characteristics (information, human capital formation, resources incl. patents/IP), and contracts – are related to risk. The price can then be expressed as:

Equation 2.1 The theory of value equation

$$\text{Price} = \text{Value}\,(\text{environment}, \text{strategy} \mid \text{institutions}) \tag{2.1}$$

Price clears when WTP = WTA (Willingness to pay = Willingness to accept) resulting in a binding contract.

The Primary Data Source for Trade Values

The (information on the) value useful for statistics is thereby tied to the *contract* where the price and terms such as which patents, which fields-of-use, the length of contract time, market access, and so on are agreed. These contracts often have a two-part tariff (fixed price + royalty).

The *reward medium* for the patent rights may range from other patents, other IP, information/data, equity, money or other assets (in which case only a "net-price" is paid out in money).

The *negotiating mechanism* of such a price is also central as the formal rules or informal norms of trade, including the possible reward mediums, directly influence the efficiency of such prices. Lacking general public markets, these prices are quintessentially private, limiting the "price disclosure" to summary data. However, this may be satisfactory for use in formulating efficient trade policy.

The theory of value equation (Equation 2.1) will now be explained for all the elements. An introduction to this dynamic trade system is first outlined. It is followed by the key economic questions of pricing risky and uncertain valued patents efficiently and how real firms manage this process, putting the *management strategies* of dealing with risk and uncertainty at the center of the theory of value. An *empirical* approach to a theory of value then follows based on a study of decision behavior of the world's leading patent licensing firms. Finally, a set of variables are proposed that will capture the data needed to calculate the value of trade flows from ideas (IP) given the strategies and institutions.

2.1 Introduction to the Dynamic Economic System of Trade in Ideas

To develop such a framework for standards and trade in ideas, it is necessary to base it on some basic understanding of the twenty-first century's *economic system.*

We cannot base it solely on existing economic theories rooted on the operation of the economic system in the nineteenth or twentieth century when commodities and services dominated the "trade mix." Today, trade in patents and other IP is increasingly *directly* (non-embodied) part of this mix (90% of company value is intangible assets of which IP is a large portion and patents the most desirable asset) and *indirectly* as embedded (embodied) in goods and services (at least 40% of goods have IP attached to them and for services it is likely much higher). IP intense firms in general are more profitable, pay higher wages than other firms. This is because the returns on human capital formation, especially through patent/IP, are higher than other investments. There are social gains to investment in patents/IP. However, it is also much more uncertain. Adding up the two "IP-components," **IP trade in the non-embedded IP *rights* and the embedded IP *related* to trade in goods and services is likely the most valuable part of international trade economy today. However, no such comprehensive statistics exist today.**

2.1.1 *Upgrading Trade Theory to Patents*

A lack of theory of value appears to be a key part in this discrepancy. At the heart of the twenty-first century, trade lies more than ever trade in ideas as a means to advance the growth of productive technology and productivity everywhere through exchange of the technology itself (for other technology, other IP (including information), funds, equity, or even work). In turn, to support a *constant* increase in productive technology – avoiding a marginal decline in the growth of the *stock of technology* – a sufficiently large increase in the *stock of knowledge* is needed and international scientific research cooperation (learning) thus feed into this highly dynamic system.

This dynamic is directly connected to an economic problem of interest since at least Adam Smith: How to extend the scope of the market?

2.1.2 *The Key Economic Question Is Sociological*

Extending the scope of the market to ideas (IP) is not only a matter of practical data collection but of a new *theory of value.* This is needed as the value of patents and IP in general is much more uncertain than for goods and services. In the current theory of value in economics, the values are known, risk free, and no uncertainty exists.

The standard treatment of risk and "uncertainty" is a game against states of *nature* (i.e., Will it rain or not in the spring? affects the price of grains in the fall), whereas in technology patents (IP) it is a game against the state-of-the-art, a genuinely uncertain process of new *human* ideas that never existed before (contrary to rain, where a probability distribution can be created based on past events).

This makes the problem a *sociological* one. To accurately measure the value of trade in ideas, a new (or modified) theory of value is required, taking into account the management of the risk and uncertainty in human ideas.

2.1.2.1 The Genuine Uncertainty in the Value of Patents

As no one can predict the inventions of tomorrow (or they would de facto have invented them) and their potential value, firms are not facing merely risk but *genuine* uncertainty in the value of their technology and patent assets used in their current and future production. Even the value of known inventions is almost impossible to predict as they can be used in radically different innovations unknow to anyone.

Just think of the vast range or application of the four-stroke combustion engine (Otto), the AC electric motor and distribution system (Tesla), the Internet (Cerf), the Smartphone (Jobs), and many others, most of which were likely not envisaged by the inventors. The inventor of electricity (DC), Volta, presented his findings to the government, and in a frustration over the inability of the officials to understand the potential of this discovery to society, he reportedly has said, "You might be able to tax it."

My favorite example is that of the computer mouse without a ball. This technology, developed by HP, apparently came from their *business* line-printers, where a camera had been developed to precisely track the location of the paper, making precision printing possible. The *consumer* side picked up on the technology through a client idea to use a camera as a mouse. When HP split up the company to sell the PC unit, the value of this patented technology for the mouse was in the billions dollar. Who at the line-printing business, or management, or anyone, could have realized the full value of such a small camera unit *ex ante*? What is impressive is how the HP *management* took a business market unit idea and built a consumer market product so successfully. This tells us that the value of goods management may in fact be underrated.

The key *economic* question is therefore how risk and uncertainty are *managed* in trade in ideas.

2.1.3 The Management of Risk and Uncertainty and How to Measure It

The analysis of risk can then be seen as a game against nature and the analysis of uncertainty as a game against people.

2.1.3.1 Uncertainty Currently Not Incorporated in the Economic Models

The game against Nature is the approach by Arrow (1964 (1953), p. 121) and Allais (1951, 1952, 1953) when it comes to assessing and pricing risk. The risk can be traded in a market as "options" on a *finite* number of states of Nature, and thereby be given a price (cost). This makes the problem of risk an *economic* one where a probability can be assessed to the outcome as a "choice Nature makes among a *finite* number of alternatives" (Debreu, 1959, p. 98). Risk is related to future events to which a probability can be estimated (such as "Will it rain in the spring?"). They are "games against Nature."

The issue of uncertainty in the sense of *genuine* uncertainty was mentioned by Knight and Keynes in the early twentieth century. Knight *separated* the issues of risk and uncertainty in his PhD thesis publication "Risk, Uncertainty and Profit" (Knight, 1921) and contends that there could be no basis for models that treat decision-makers as having "correct models" with which to make decision, i.e., the *management* problem. Keynes added the "absence of coordination," i.e., the *market* (or institutional) problem. Since almost 100 years this development has essentially stood still at these early contributions and remains at stating the fact that uncertainty is not a matter of risk and markets may coordinate risky but not uncertain exchange. Finally, to sum up the state of the economic art in dealing with uncertainty, Phelps wrote in an Op-Ed (2019) that "uncertainty is not normally incorporated into our models" and that "the uncertainty revolution has not yet succeeded."

2.1.3.2 The Social Process of Managing Inventions

The discussion of option pricing does not include an exchange of messages – communications, negotiations – between people trading the risk. Only the *outcome* is discussed. This is because nature obeys laws whose outcome can be assessed. The *social process* of negotiation preceding such prices is ignored.

When it comes to inventions, which can be defined as solutions to a technical problem formulated by the creative minds of learned people, these *human ideas* never existed before (if not invented by another human somewhere else or forgotten through time). This is thus what makes the problem of uncertainty (in the value of inventions) a *sociological* one: How then to resolve the uncertainty in the value of current and future inventions made by humans?

The process by which ideas come about is a highly dynamic one where typically individuals or small teams face a technical problem at hand, trying to solve it. Since one cannot study the ideas which have not yet been created without creating them first, the value of new ideas is *genuinely uncertain*.

The process between people who will invent the technologies in the future therefore needs to take into account this uncertainty in their own investment decisions. This is thus a *management* problem of which *technologies* (technology areas) to invest in (to create a competitive advantage) and which ideas to *further* invest in to develop a new, inventive, and useful, i.e., patentable,[1] technology.

Genuine uncertainty can be likened by a man throwing darts at a dartboard trying to hit the bullseye, a complex *learning process* where new throws require *creative* ideas, often in an environment with friends giving advice.

With nature it is different. Here "laws of nature" dictate its behavior and one can study Nature that exists.[2]

2.1.3.3 Patents as Production Assets Establish Strategies

The social process of reducing uncertainty is captured in the sustained strategies among several strategic options. For firms, strategy is essential as they only have limited resources and need to choose where to invest their funds to be most successful given what all others choose. The novel dimension of the theory of value is thus the need to include the patents as production *assets*, not only their cost or contracting cost.

For statistical purposes, both *transactions* and *assets* are needed to be measured, and both are new for international financial accounting standards and balance of payment statistics when it comes to ideas.

They are simply not covered in a *comprehensive* way in neither financial accounting standards, statistics, or international economic law.

A long fruitful road lies ahead!

2.1.4 Theory of Value of a Dynamic Economic System with Uncertain Assets

To analyze the social process preceding pricing and exchange, the *decision behavior* – the management's choice of strategy – of trade-in-ideas motivated inventors, investors, innovators, and university and institute researchers needs to

---

[1] The patentability criteria in the US system are: new, nonobvious, and useful, and in the European and ROW: new, inventive step, and industrially applicable.

[2] Even determinism, which Einstein supported to the end, is on its way back after relativism, thanks to the black hole information paradox studied by Hawkins. This makes the problem of inventions a very interesting one, as some humans – the inventors – can *create some new use* of the "forces of nature" out of "nothing" except a mere understanding of them and how they can help solving a technical problem. The view of human nature is here as a *creative individual*, not merely a "mechanistic" deduction machine. It rejects the "mechanistic brain" approach by asking "why?" one more time, as a personal communication with a Jewish professor from Stanford revealed: "God created us with a brain so we could *ameliorate* on his creation, nature." This is the de facto insight of the second industrial revolution: the integration of science and technology and the reason *why* technology exists. It is a human *creation*.

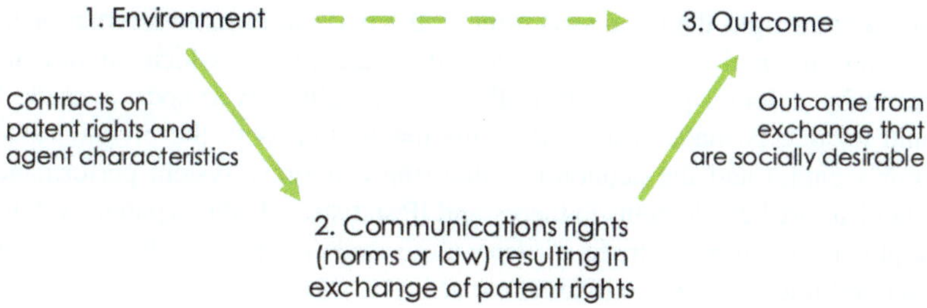

Fig. 2.1 The rules – whether norms or in law – negotiated by members of and enforced by the institutions embodying them should give incentives leading to a certain decision behavior (strategies) which in turn is socially desirable

be characterized. Such decision behavior has important application to the development and verification of theories, policies, and practices of the *economic system.*

An analysis of the economic system is first singling out the management strategies (behavior) as a separate element in the economic system. It is followed by an *empirical* characterization of strategies based on a study on what patent licensing active firms *do* as they attempt to resolve the uncertainty in the value of patent rights. The resulting trade is then the trade in ideas that needs to be measured to capture the trade flows.

For these purposes, the economic system is divided into two distinct parts:[3] (1) an *economic environment* consisting of the agents and agent characteristics including their strategy and the contracts traded, and (2) an *institutional environment* defining the rules of the property rights under which agents may (a) get *ownership rights* granted on ideas by applying for and enforcing patent rights, (b) *communicate* the social process of finding buyers/sellers/traders, negotiating and auditing a contract, and other "messages" between agents culminating in exchange, and (c) *exchange* the contracts for the purpose of modifying their patent rights holdings and obligations in line with the (investment) *strategy* they pursue. Since such contracting on patent licenses or transfers requires negotiation between the parties prior to exchange the rules of communication rights are as *important* as the rules of ownership rights on the ideas (the patents) (see Fig. 2.1).

In trade in ideas, there are therefore two "exchanges" taking place. First, with the government where *disclosure* of the invention and paying fees are "exchanged" for a patent right. The private ownership right of the invented idea is in fact two separate rights: a right to *exclude* others from using the invention without consent, and a right to *transfer and license* the patent right with mutual consent, creating the *market* in ideas. Second, exchanging the right between inventors (holders), traders (who buy and sell rights), and innovators (users) to allocate the rights in the economic system where the private and economic value is the highest, using transfers and license agreements.

---

[3]This analytical approach is based on the institutional approach to microeconomic systems outlined in Smith (1982).

This second right then connects directly the trade in ideas to international trade policy and international economic law, the outcome of which should also be measured by a social gain, such as Pareto optimality (everybody is made better off, and no one is made worse off). This means that both the private value (the inventor's gains) and the economic value (the economic system performance) of trade in ideas are key elements in theory and IP statistics. Today's patent systems and the sophisticated use by trade-in-ideas motivated inventors will now be briefly summarized under these two distinct parts.

1. The contracts used in the real **economic environment** are both complex in the way they are negotiated and simple in their "types." For the purposes of trade in ideas these are two types, which come right out of the patent rights: transfers and licensing contracts.

   What make them complex is the assets involved in clearing the contracts, i.e., the payments for the rights exchanged. In essence, a large portion of the contracts are "barter contracts" where pools of patent rights are exchanged for other patents, other IP like data (information), trademarks, designs, etc., equity, and even work. Indeed, this decision behavior – or strategy – indicates the needs of firms to maximize their *production assets* of which technology is at the heart, both for the firms' operations and the operation of the economic system at large. In this investment strategy, they leverage existing assets, and money in a complex exchange of and for "technology rights." Other contracts are "pay only" contacts, yet other transfer contracts. These payments can be in "any" payment system: official currency, in cyber currency, or "tokens," which may not be tradable or tradable for legal tenders. Transfers are also done to create common pools of core technology, but the holders then have a share or a license with the "common pool" of production assets. Finally, these pools can also be "securitized," i.e., sold to a special purpose vehicle (SPV) who owns the revenue streams from licenses, in a way to "monetize" the value of patent licenses and capitalize inventors or fund further investments.

   The most common contracts today are probably cross-licensing contracts where (typically very large) patent portfolios are shared between the agents and the "clearing" of their respective relative value is done on the margin, based on a few uniquely valuable patents on both sides. This results in a net payment to the agent with the most valuable patents. These contracts are often limited to use in certain goods and services, certain standards (such as in telecom FRAND agreements). Thus, what is contracted can be seen as "production assets" for certain *technology areas* for certain *field of use*, *time*, and *geography* (markets where the patents are validated) against the payment for this technology in production assets, including financial assets, which may be the smaller part.

   The agents and their characteristic are important as some are inventor-only, trader-only, inventor-innovators, or simply innovators using the technology for new goods and services. These agents are then characterized by highly *specialized knowledge* in certain technology areas driven by gains from specialization through exchange. Such "industrial organization" or structure, or cluster maps are key to understand the dynamics of the trade in ideas market, to assess where the gains form trade in ideas are achieved through specialization. This is Adam Smith upgraded for the twenty-first century trade in ideas.

Exchange → Specialization in technical knowledge → Productivity gains

The stock of technology and the stock of (specialized) technical knowledge (useful for inventing) are then key elements of a socially desirable outcome in a trade in ideas theory of value.

Contract *types*, agent *types*, including *legal entity type* (which may have limitations on some international transactions), agent characteristics such as *strategy* choice, *knowledge*, knowledge *search* limits such as access to human capital formation through hires or *integration between technology and science*, *information search* possibilities (digitalization, AI) are all characterization of decision behavior impacting value from trade in ideas and thus need to be included in a theory of value from a business management and an economic environment perspective.

2. The **institutional environment**. Real trade in ideas (IP) involves many *different* institutions to define the property rights of (a) communication in the patenting and trading mechanisms for ideas (IP), and (b) ownership of the ideas (IP) that are later to be traded. In real markets these institutional arrangements are complex both for communication and ownership rights.

   (a) *Communication rights* can be cultural norms or formalized such as in a free trade agreement.

   For example, in Chile, firms do not have the "culture" of paying university researchers for inventions as they are considered "teaching institutions," the firms. Many countries also do not allow international sale or licensing of patents based on government-funded research.

   In Israel you can license technology developed based on research from government grants but only after you pay a fee for doing so. Such intricate communication rights – whether based on cultural norms or formal trade rules – thus create incentives for a decision behavior affecting the trade-motivation of agents.

   Communications right for trade in ideas is still a rather unregulated area. Cultural norms on transferring "traditional knowledge" are key elements for many countries.

   (b) *Ownership rights* are based on international agreements (TRIPS, Paris Convention, EPC however limited to "European" countries as defined in 1960, including Turkey, a non-European country), regional agreements (EPC, other), national patent laws, and ultimately implemented by a national patent system through national (or regional) patent offices and courts. Filings can be done in a "top-down" fashion by the international PCT system, or "route," with a PCT authority and subsequent "regional route" or "national route" for examination (where used) after 18 months or in a "bottom-up" approach with national first-filings under the Paris Convention of 1883 and then granting of the rights at national or regional levels or a bundle of national patent under the EPC for "Europe," just to barely scratch the surface. Not to mention the opposition procedures of EPC and now also the United States, automatic injunctions when infringing like in Germany and "similarly" in China or practically no injunctions in the United States, but only since the eBay case. There are also key issues of patentability, exhaustion of rights (typically when a good is sold the first time, but not for software, which are often a bundle of components from many developers), carveout for pharmaceutical patent, and how to deal with traditional knowledge. The patent office is the authority administering the patent (IP) conventions in question.

As the IP systems are quintessentially national systems, national law defines the rules of a patent (IP) and courts to enforce the patents if infringed. The patent courts, which may be administrative inside the patent office for immediate post-grant "opposition" procedures or (mostly) part of the national civil court system (together with all other civil lawsuits) or even the criminal court system (in some countries infringement is a criminal offence) or in the best of world's specialized patent courts that exist in main patenting countries such as the United States, Germany, and China, or a combination of all of the above. Together they form the patent system in practice.

Today's system is really made by and for the legal experts, not so much focusing on the economic gains from specialization which trade-motivated inventors can realize. Major reform is needed to restore the patent system for individual and small team inventors where most inventions take place.

Together with the *communication* rights (including cultural norms, regional and international agreements) and *exchange* rights (right to own, rights to demand payment or transfer and license), the patent system (patent office and courts) becomes a *trade system* for ideas.

2.1.4.1 The Theory of Value Summarized

The trade value is thus conditional to the *actual operation* of the economic system to reduce risk and uncertainty, i.e., the agent characteristics, including strategy (decision behavior) *and* the institutional arrangements (communication and ownership rights). The value can therefore be summarized as:

$$\text{Value} = f\ (\text{environment, including strategy} \mid \text{institutional arrangements}) \quad (2.2)$$

The environment: agents and agent characteristics incl. strategy and contracts
The institutional arrangements: communication and exchange rights

The value is thus dependent not only on the competitiveness of the environment – number of specialized firms, size of market – but of the strategies inventor firms use, which in turn, is conditional of the institutional arrangements for communication and ownership rights. These strategies will be outlined in the next section.

2.1.5 An Empirical Model of Value Based on the Strategies Firms Use to Solve the Uncertainty Problem

To develop a theory of value for uncertain patent rights, we need to characterize the (a) *strategies* and other relevant firm choices – their real decision behavior – given the (b) *institutional* arrangements de facto in place ("in practice"). The (c) *variables* in the resulting model will be listed at the end, defining the data needed. The strategies to reduce uncertainty may be conditional on the institutional arrangements, thus both business policy and government's institutional policy impact – through mainly company culture and international economic law (customary law, treaties, agreement, standards) – how the uncertainty in value is resolved and then which patent rights are traded.

How then to determine this value? The approach taken is a study of what patent licensing active firms actually *do* to manage the uncertainty and *why* they do it. Then characterize those actions and motivations in terms related to the economic system described, which in turn can be the *variables* of such a model of value.

A normative theoretical approach can be developed once we have data on this global trade, provided, in part, by the new proposed IP statistics.[4]

2.2 The Strategies Outlined

The *empirical* results a study of strategies will now follow to classify and characterize the strategies. Four "types" of strategies were identified in a study of the most patent licensing active firms in the world in 2011–2013. This study was reported in Ullberg (2015a, b).

2.2.1 The Strategies Which Resolve Uncertainty

The strategies share the common theme of creating *mutually assured self-restraint* (MUST) not to sue or otherwise harm the other party when new competitive technology is invented by either party.

This *decision behavior resolves the uncertainty* in the future value, making the value of the patent right less uncertain and more tradable (see Table 2.1).

---

[4]For an initial elaboration on a rational approach to price theory for how firms manage risk and uncertainty in markets based on experimental data, see Ullberg (2012).

Table 2.1 Strategies firms use to create trust in each other's actions

Strategy 1: Staying clear
In this strategy, two large firms (or countries) compete head-on in the product and service markets using patented technology they developed (or bought) themselves, but they do not license it. To respond to a threat that the competitor will get "too close" to their core technology with their patents, and thus weaken their technology advantage, both patent close to each other's core technology areas, slightly overlapping each other's patent portfolios. This means that if one will sue the other, the other will sue back, both losing out. So, they trust in each other's actions, that they will not sue each other because of the destructive costs and thus "stay clear" of using each other's "core technologies." This appears to be exactly the "mutually assured destruction" strategy (MAD) used in the cold war between the United States and Russia. To use such a strategy, 1000s or 10,000s of patents are often needed. This is the game for the Kings of patenting.
Strategy 2: Strategic alignment
In industries where there is a high level of interoperability required, such as telecom, electronics, or software, where standards may also be used that include patented technology, the opposite of Strategy 1 is used. A binding contract is established where the whole patent portfolio, or whole slices of the portfolio, are mutually shared between the trading partners. In addition, and this is key, all *future* patented technology that reads on the shared field (or standard) is also shared, for the duration of the contract. These contracts are called "capture period" contracts as each party will "capture" the new technology developed by the other. The price is often the net-payment, having considered each other's portfolio strengths. They typically are 3–5 years long and can be of two types: "guillotine" contract where all right to use the patents end at the end of the contract or "patent life," which allow the partners to use the technology for the duration of the legal life of the patent. So, they trust in each other's actions because they have *waived the right to sue each other for a time* in some areas of technology exchange. The strategy allows for "strategic alignment" in some areas and for a limited time. Both parties gain from the specialized knowledge of each other. This strategy can be used with far less patents, 100s or 1000s may be needed. Such cross-licensing is probably where the majority of the transactions take place today.
Strategy 3: Marginal contracting
This contract is the only contract that is covered by neoclassical economic theory. Parties purchase or license patented technology "on the margin," complementing their portfolio. However, these patents are typically tested out in a market where their value has been demonstrated, thus the risk and uncertainty in the value is low in this case. It is close to the neoclassical approach where the "cost reduction" is of interests as price = marginal cost. However, such cost-reducing patents are essentially process patents and not product patents, which may increase the revenues in a non-marginal-cost way. "Marginal contracting" refers to the "marginal value" added to their portfolio, not only a possible "marginal cost" reduction. In recent years this licensing has been growing, in particular in relation to venture capital who often demand patents as assets for their investment. They trust each other's action by being able to *sue in a court that honors the inventor* (rule of law applies), would someone infringe their rights, because the value is more certain. Perhaps 10–100 patents are needed here. An important ownership right here is an *injunction*, which have been drastically weakened since 15 years in the United States, making it difficult to pursue this strategy for individuals or small teams, especially in developing countries. Much policy discussions are on this strategy, whereas strategy 1 and strategy 2 appear to be the more common.

(continued)

Table 2.1 (continued)

Strategy 4: Systemic abuse
In this strategy, firms buy up low-value patents or any patents that companies of consumers may infringe on and then threaten to sue them in court. This is particularly a big problem in the United States because of the "discovery" procedure. In which the plaintiff can have a judge demand private information from the defendant, which is typically a very costly procedure. The suing firm then offers a license at a price lower than clearing the alleged infringement through the courts. The countermeasure here is to go to court (costs about $1m or so), then put up another $10m or so to have the patent cancelled by the patent office. Since each party pays their cost (in the United States), the defendant "runs the plaintiff out of business" as there will be low penalties to pay for a low-value patent, if it even holds up in court. So, they trust in each other's action by threatening to *running the plaintiff out of business* (at initially very high costs for the defendant). The next time around the "systemic abuse" by the plaintiff firms comes back to the defendant firm with valuable patents instead, moving to strategy 3.

Strategy types based on the empirical study :

1.staying clear, 2.strategic alignment, 3.marginal contracting, 4.systemic abuse (2.3)

2.2.1.1 The Strategies of Mutually Assured Self-Restraint ("MUST")

The common for these strategies, or strategy options, is that the parties attempt to create situations where they can trust in each other's *actions* not to sue or act opportunistically. **These strategies thus create a *mutually assured self-restraint* (MUST) and are the essential component in the empirical theory of value as they have the purpose of reducing the uncertainty in the trade value to a level where only risks remain for which a probability distribution can be created.**

2.2.2 *Reward Structure by Strategy*

Given each strategy's *reward structure*, a price is negotiated between the parties in the strategies where transactions take place between parties. If transactions take place within a hierarchy, and transfer prices may be used, these need to be determined by comparisons to known "market prices." If transactions take place as a result of litigation, the out-of-court settlements reflect cost and if in-court settlements is the outcome the costs are stipulated by law. These prices reflect the value a byer is willing to pay, WTP, and the offer a seller is willing to accept, WTA (see Table 2.2).

The price is negotiated using the reward structure that is the highest for the given technology.

Table 2.2 Reward structures for the strategies discovered

Strategy 1, *staying clear*, will yield gains from specialization between firms as they focus on different core technologies, which may yield different profits in the goods and services markets. *Transfer prices* between patent holding entity and producing entity indicate the market value of the IP.
Strategy 2, *strategic alignment*, will yield gains from exchange through technology producer and use surplus. A net-value between cross-licensed portfolios is contracted, reflecting the relative price of the portfolios. The paid net-price indicates the additional leverage of the firm's specialized human capital formation (knowledge) turned into patents.
Strategy 3, *marginal contracting*, will yield gains from highly specialized exchange through increased surplus.
Strategy 4, *systemic abuse*, will reward abuse of the system by asymmetric cost of defense of rights, a rent-seeking behavior, if a strategic response is made to run the abuser out of business. The price paid in settlements is an indication of enforcement costs needed to be paid in order to have a functioning market using strategy 2.

$$\begin{aligned}\text{Price} = {} & \text{Value, given the reward structure of each strategy option(firm behavior),}\\ & \text{given institutions (constraints which structure economic,}\\ & \text{social, and political interaction)}\end{aligned} \tag{2.4}$$

2.2.3 The User's Values and Two-Tiered Prices

Introducing the patent system does allow the inventor to exclude, license, and transfer the protected technology with mutual consent in exchange for teaching the world about it plus paying a fee to the government. The potential *user* then has a choice to make whether the patented technology is worth paying for, i.e., the value of its *use* comes into the price equation (not its "marginal cost"). The value is not based on the value of a product bought by a *consumer* (in a consumer market) but through a license acquired by a *producer* (in a producer market). This suggests that in order to better understand these patent markets, price should equal the value for the user(s) of the idea. Such licensing creates a (highly) competitive market in technology where prices really have to match value of the *user's products*.

The neoclassical concept of marginal cost pricing is therefore not applicable to patent licensing, which instead tend to be the result of a stream of effort that can be equated to *producer value*. In those environments and because individual bargaining (with private values) is more realistic, a scheme to expropriate consumer surplus (such as price differentiation) is more apt to a producer market since this may achieve the most efficient outcome. Neither would this necessarily generate an economic profit for the patent owner since consumer valuations are not necessarily significant and would be hard (if not impossible) for the patent owner to predict ex ante (Ullberg, 2019).

The value may thus not theoretically be zero (as in marginal cost for an additional patent license), but rather what people are *willing to pay* for the contract, which is the value (less risk and transaction costs) that users are able to realize *after* the genuine uncertainty associated with the patents is managed by one of the four strategies.

It is not necessarily monopolistic competition either (as in average cost). In monopoly theory, the monopoly is an institution-less process where the implicit assumption is that buyers reveal 100% of demand and sellers hold back supply (Smith, 1982, p. 946, Proposition 8, footnote 28). Only take-it-or-leave-it kind of offers (posted-offer) may result in monopoly prices. However, the buyers cannot reveal a demand they themselves do not fully know. It is only in repeated experimenting in the goods and services market that the value (demand) can be fully known, if ever. Sellers, on the other hand, typically try to sell their technology to the broadest use, which is socially preferable as the "distribution" of the technology is then broader. In demand-side bidding experiments, we also see that buyers do not bid to zero but bid up to the second highest value (the closest competition) where there are competing buyers (Ullberg, 2012, Ch 3).

This producer value dynamics results in a *two-tiered tariff* in many cases. A fixed fee (annual or one time) plus a royalty (on revenues). The up-front paid fixed fee is used for *transfer* of some risk (both for transfers of patents and licenses) and the royalty to *share* the risk by paying a fraction of *realized* revenues once they are known, i.e., the risk is gone. These payments are, however, *conditional* to the strategy used to reduce the uncertainty in the value. The payment for the user's value is thus done up-front *and* ad hoc. The ad hoc royalty amount paid, "the price," is thus dependent on the price of the product or service it is used in, i.e., how much value it can add for the final good.

2.2.4 A Socially-Strategic Equilibrium

A final and concluding remark on the strategies dealing with uncertain values is their relation to strategies with certain values. The presented strategies do not appear to be Nash Equilibriums (Nash, 1950), as the values of future technology cannot be known but are genuinely uncertain, and we are interested in a *cooperative* equilibrium. It is a *socially-strategic equilibrium* built on trust in each other's actions motivated by *mutual self-restraint* created by discrete investments in select patent *assets*. The *asset* side (the balance sheet) must be included. They foster cooperation by creating conditions of mutual self-restraint, which in turn creates trust in each other's actions not to sue or otherwise harm each other. The uncertainty is reduced and the patents desirable to exchange and license can now take place at merely a level of risk as the inventors will not harm each other (behave opportunistically) when they come up with new, more valuable inventions. In fact, the reduction of uncertainty stimulates the creation of more ideas as the upside is not a potential trade value, long-term specialization, and promises of gains from trade.

The Nash Equilibrium is a solution concept to a situation where agents cannot make binding cooperative agreements. In this case the agents do not make any binding agreements based on cost (which is unknown) nor when it comes to trust but they adopt an investment strategy in which the patent assets create a situation where both parties will lose if infringing each other's future rights of unknown value (cost to the other). Such a harmful solution is a Nash Equilibrium solution when the inventions are known and their value.

Importantly, this social process is not only "sustained," as in Nash's solution concept, but appears to create some common expectations, which in turn are supported by some *theory* or at least *rule of thumb*. Most commonly, the profit from new ideas is valued ad hoc at 22–28% or "rule of thumb" in patent *licensing* of 25% of profits.

Aumann proposes in "Assessing Strategic Risk" (Aumann and Dreze, 2009, p. 2) that games against nature and against people can be *equated*, i.e., reduced to a risk. Even if risks against nature and risks against people can be equated, when it comes to uncertainty they are different and needs to be treated *differently* (as Knight suggested). Aumann's proposal appears to suggest that when it comes to risk, people behave like nature, i.e., according to some "laws of nature."

This indicates that game theory with known values is currently out of reach of the uncertainty characterized by trade in ideas. This proposition is further enhanced in Holt and Roth (2004, p. 4002) "The Nash Equilibrium: A perspective," where the need to incorporate "strategic behavior" is mentioned as an area of future necessary expansion of game theory.

> "...the challenges facing game theorists include learning to incorporate more varied and realistic models of individual behavior into the study of *strategic behavior* and learning to better use analytical, experimental and computational tools in concert to deal with *complex strategic environments*" (italics added).

Trade in ideas is clearly a "complex strategic environment."

The economic system thus operates in a way different to both the neoclassical approach (where there are *no* risks) and the Nash Equilibrium (which covers strategies in noncooperative games where information about the outcomes is *known*) and the game theory approach in general (which needs to include behavior in complex environments).

Under uncertainty and risk, the solution requires not only knowledge about the costs of others' options but also a strategic equilibrium of mutual self-restraint created by investments in patent bundles. This solution concept of mutual self-restraint thus requires both the investment in strategic assets (the balance sheet) and operations (the income statement).

If trading ideas is an option, then this is an investment question (balance sheet), not a cost question (income statement). The patent/IP *assets* therefore need to be taken into account when analyzing the trade in ideas with *uncertain* assets. This has impacts on the financial accounting standards, which currently have a very limited recognition of patent/IP as assets.

One therefore must take into account the asset side of the agents' patent/IP assets. The statistics therefore need to be based on assets rather than transactions, as some transactions have the purpose of acquiring productive assets to resolve uncertainty.

As a matter of fact, most such interments are used to manage uncertainty (and risk). Only 2–3% of granted patents are used in any good or service. Perhaps 50% are used for blocking others from using them in an attempt to protect current technology and create options for future technology. Of the remaining 48%, many are not economical right now or have not found investments to be developed or bundled with for a useful technology or are currently used in a competition, competing for a market use in this dynamic system.

In summary, the price = value, which is a function of the competitiveness of the environment, the completeness of the contracts, the strategies chosen by a certain investment strategy, and the institutions. The prices are often conditional to the performance of the licensor over time (royalty), not only a "one price" (fixed fee).

Data on all these *elements* need to be collected to understand the performance of the economic system given its economic environment and business strategy.

2.2.5 Non-satiation Postulate for Uncertain Values

According to the postulate of *non-satiation* in economics, an *autonomous* individual given a *costless* choice always chooses the alternative that gives more reward medium (utility is a monotone increasing function of the reward) (Smith, 1973, pp. 22–23). For values under uncertainty, this implies that the *strategy* used to reduce uncertainty with the highest reward is chosen (which depends on cost of capital, economies of scale, inventive capabilities, etc.).

In the case of trade in ideas, the "reward medium" includes money and/or patent or patent contract *assets* (giving the right to use the patented technology), or other assets such as information, technical assistance contracts, or equity.[5] The impersonal trade choices (between inventors and innovators) thus compete with the personal trade choices (invention and innovation in the same hierarchy), making economic

[5] If a person is hired as a result of an invention, and the patent rigth is transferred, then that person will receive a wage. This is different from a "payment" in terms of a temporary technical assistance contract. A distinction need to be made between IP-related markets and labor market transactions. The latter contract should be recoreded like a merger and acqisition transaction. This distinction is important when understanding the rationale for economic organization and corresponds to the concept of specialization between inventor and innovator, inventor and researcher, inventor and manager, and inventor and investor. The implication of this specialization for financial accounting will be discussed in a later section as financial accounting standards have their roots in the *specialization of function* between a "Schumpeterian" entrepreneurial investor and manager (1912), and later a "Schumpeterian" intrapreneur and manager (1934).

organization a key issue of the problem in institutional performance. If the behavioral properties of the institutions are such that market exchange is preferred to hierarchies, then the gains from specialization are likely to be higher.

The value of a transaction is related to the value that can be created in the future by shifting from a low-reward strategy to a high-reward strategy. This shift is made possible through the accumulation of patent rights, which are necessary to create the desirable mutually assured self-restraint.

Example 2.1 If a firm has the choice between a marginal contract for cash and a cross-licensing agreement including attractive patents, they may choose the cross-licensing in order to pursue a strategic alignment strategy in the next step. An example here is the "Rockstar" patent acquisition consortium used to buy the Nortel patents in the bankruptcy auction in 2011. The leading and competing telecom giants/users – Ericsson, Apple, Blackberry, Microsoft, and Sony – created a "bid company" to outbid in particular Google, effectively hindering them from entering the telecom market. The patents were pooled and licensed to the consortium. The price was $4.5b. They were then accused of playing Strategy 4 by attacking Google, Huawei, Samsung on patent infringements. This was settled in 2014. Also "non-strategic" patents were sold for *defensive* purposes to RPX Corporation, as "trader" attempting to play Strategy 3. The example shows that huge sums are being invested in portfolios to guarantee that the reward structure of the strategy is sustained.

Similar examples can be found in the agreements on standard essential patents, SEP, traded under FRAND terms, marginal contracting by smaller inventors to large companies of key technologies, and also, due to institutional weaknesses, abuse by patent holders creating holdout situations under the threat of prohibitive court costs.

$$\begin{aligned}\text{Value} = \text{Max}(\text{reward medium}|\text{strategy}), &\text{ convergence time to an equilibrium price} \\ &\text{may take many years of investments, transactions, and M\&A}\end{aligned} \tag{2.5}$$

2.2.6 *Investment Strategy and Maximizing Value*

This price optimization postulate leads us to the final point of strategy *implementation*: an investment strategy to create a patent position that can be used in an "optimal" reward structure. The management of uncertainty can then be reduced to the *investment strategies* enabling the strategic positions that create such mutual self-restraint. This, in turn, is a matter of the cost of capital of inventor firms. A low cost of capital would lead to a more investments, more trade, and more gains from trade and a high cost of capital the opposite. As the cost of capital is directly influenced by taxation policy, the implications are also in this second policy area as well as institutional policy.

Table 2.3 Investment needs to play the different strategies

Requirements of patent portfolio and entry investments by strategy		
Strategy	**Patent portfolio**	**Investments**
Strategy 1, staying clear	**>10,000 patents**	**$10b**
Strategy 2, strategic alignment	**1000–10,000 patents**	**$1b–$10b**
Strategy 3, marginal contracting	**100–1000 patents**	**$100m–$1b**
Strategy 4, systemic abuse	**1–10 patent**	**$1m–$10m or less**

The average cost of a patent is here assumed to be $1m over its lifetime. See specialist literature and newsletters for other estimates

A special low-cost-of-capital legal entity is proposed as a solution in this area. A 0% (or negative) tax is granted in exchange for higher capitalization, thus shifting some capital investments from innovations to inventions in a "risk neutral" way (see Table 2.3).

The decision behavior of a firm is thus dependent on the reward structure desired, which creates very different investments needs, which in turn are determined by, in part, the *cost of capital* of the firm. This means that not only the revenue side of the firm but also the asset side of the firm are necessary in determining the price under uncertain values.

$$\begin{aligned} \text{Value} = \text{Max}(&\text{reward medium, discount rate}|\text{strategy}),\\ &\text{convergence time to an equilibrium price may take years}\\ &\text{of investments, transactions, and M\&A} \end{aligned} \tag{2.6}$$

2.2.7 *Industrial Organization*

There are no considerations with respect to industrial organization, as inventions most commonly are made locally by individuals or small teams. It is important that the smallest legal unit can be considered an agent. The agent characteristics may include size, foreign/national ownership, group hierarchy, and so on as a matter to identify which agents have the most inventive individuals and teams. As most companies are small, and most patenting takes place in small companies, the focus is on the value of the idea, the IP *right*, not primarily the value of the agent owning them.

2.2.8 *Concentration of Capital*

Of particular importance is to avoid concentration of inventive activity to a few firms as such concentration may stifle inventive activities. One should also avoid state capitalist solutions that result in the same concentration. Dynamism that promotes new ideas and new inventive firms are needed. An important measure is then *new technology/same technology* used in products and services. Joan Robinson's critique of the market system in "What are the questions?" (1977) was in fact that the capitalist system led to a concentration of capital.

However, this critique does not appear to take into account the human capital and new ideas, which creates a more dynamic system where such concentration does not take place as new firms can enter the market and attracting funding. A critical measure is therefore *allocation of capital* as: "*more of the new*"/"*more of the same*." This is directly related to the efficiency of the market in patents (IP), serving as the selection mechanism for which ideas to further invest in wherever they are from developed or developing nations' individual/small teams, SME, large firms/ MNC or universities/institutes.

2.2.9 *Integration of Science and Technology*

The human capital formation to which firms have access is determined by the hiring of highly educated individuals, joint ventures with other firms, and by the cooperation agreements with research universities and institutes. The patent ownership and rewards play a key role in this cooperation. Sometimes a special firm is created where the IP rights are owned but the research takes place at the basic research institute or the inventors offer their patents for sale, or the research institutions offer them for license to firms. The output of new technology at a *constant* rate is thus conditional upon such access to new knowledge from research institutes. Such access and "flow of new knowledge," i.e., continuous *learning*, is instrumental to the value of future technology. A higher flow of new knowledge enables a constant flow of IP, *maintaining and increasing the value* of Strategies 1, 2, and 3. Knowledge is a characteristic of agents, and the successful integration of science and technology is conditional to research and commercialization policy of both firms and universities/institutes.

$$\text{Value} = f\ (\text{environment, learning, strategy} \mid \text{institutional arrangements}) \tag{2.7}$$

2.3 The Institutions

In a similar way to the characterization of the strategies, the institutional arrangements can be characterized.

2.3.1 Institutional Performance and Strategy

The institutions are characterized by a classification of cultural norms and formal agreements. This classification is then an independent variable of the value (which depends on the strategies, which in turn depend on the institutions). As these institutions are made up of many different institutions in many countries, economic trade areas, or multilateral ones, the actual treaties – and their implementation – should be part of the characterization. Here several institutional characterizations and rankings exist in key areas. There are also indices, for example, to assess institutional quality such as for the patent system.

In addition, *special indices* for trade in ideas may be the best way forward, summarizing the propensity to trade in ideas – trade-in-ideas motivated inventors – at a country or economic cooperation area level. Such indices would then be possible to use to calculate "regional prices" and provide the strategic and institutional explanation for divergence in prices between regions, a formidable policy instrument lacking today.

In addition, the actual regional and general trade agreements have names, and their implementation is closely followed and reviewed. For example, the WTO holds biannual "trade reviews" of countries to estimate compliance. Such institutional "rankings" could be used in the evaluation of which strategies are used, conditional to the institutions in place. The institutional performance and behavioral properties of trade in ideas could then be measured by the incentive the institutional rules have on the decision behavior of firms in terms of their strategy.

What is desirable is an efficient market with a behavioral property of a strategy that has the highest rewards, i.e., the more *cooperative* Strategy 2 and Strategy 3.

Once the strategies are characterized for the agents, the value can then be related to a particular institutional trade policy. Such a policy can be developed informed by the institutional performance and behavioral properties of the different markets in patents. This and other theses can be tested with the new statistics at hand.

The institutional rules creating efficient markets in IP need to be better understood or practiced and the IP statistics is therefore instrumental in informing such a policy on a field-experimental basis.

2.3.2 Price Equals Value

Prices that are paid for the rights are *negotiated* between parties. This is a process captured in the *classical* economic approach of *price making* agents (not *price taking* as in neoclassical theory) where the willingness to pay, WTP (bid) and the willingness to accept, WTA (ask) clear the market (see Equation 2.1). The markets clear when WTA = WTP. It is postulated that such an agreed price = value, which determines the value (see Equation 2.2).

Equation 2.2 Markets clearing conditions

$$\text{Price} = f\,(\text{WTP} = \text{WTA}, \text{environment}, \text{strategy} \mid \text{Institutions})$$

Patent assets sustain a certain strategic behavior to reduce the uncertainty to tradable risks and specialized knowledge is integrated into the price.

If learning is added, then the access to the scientific community is integrated (see Equation 2.3). The education policy gives incentives to such learning.

Equation 2.3 Learning sustains the growth in technology and is priced

$$\text{Price} = f\,(\text{WTP} = \text{WTA}, \text{environment}, [\text{learning}], \text{strategy} \mid \text{Institutions})$$

Patent assets (technology) sustain a certain strategic behavior to reduce the uncertainty to tradable risks and constant growth in specialized knowledge (science) is integrated into the price.

This result is the basis for what data is needed to be collected and what methods are needed to be developed, in addition to current data collection and summaries.

2.4 List of Variables in the Empirical Model and Their Meaning

The variables discussed in the proposed empirical theory are now summarized under headings. It follows the economic systems analysis and includes learning and is extended for agent's strategies to resolve uncertainty in value (see Fig. 2.1 and Equation 2.3).

2.4.1 Economic Environment

The economic environment consists of the contracts traded and agent characteristics. The contract is especially important as it is the basic source for financial accounting of firms. All transactions start with a contract (formal or informal). The agent characteristics are also important as they reflect the de facto "industrial" structure of the economic system. The environment is defined by the collection of characteristics of all agents.

Environment	Characteristics
Contract	None, transfer, licensing, cross-licensing; price/tariff (fixed and/or royalty); reward medium (money incl. NFTs, IP, data, other); field-of-use (IPC), market coverage (country), and length of contract (years or "guillotine," patent life), re-tradable or not.
Resources	R&D capability. This is the human capital that is used to develop new technology, use it in prototyping.
Capital (cost of capital)	The investments – financial capital – define the fixed environment (short term), which specify the limitations and research opportunities for changing strategies, tastes, knowledge, and technology. The cost of capital is thus essential to the environment. This has bearing on savings, taxation, and FDI policy.
Agent behavior	The most central dimension are the strategies used to reduce uncertainty.
Agent ID	Legal entity identification: inventor, team, firm, university/institute, other legal entities including new ones (and their limitations on transactions, capital requirements, and taxation). This will distinguish the producers of technology from the owners and the users (or blockers). The different legal entities define, in turn, "tastes" for and "utility" of technology and knowledge.
Strategy	Type of strategy pursued to reduce uncertainty and trade. The four proposed strategies will be the basis implying the investment strategy of agents, i.e., their allocation of resources for invention and patent right ownership.
Investments	Investments made in patent assets (own developed, purchased, licensed)
IP assets and liabilities	The production possibility frontier (PPF) of agents in terms of protected technology. Who produces, owns, and uses the assets. This indicates the stock of technology in the world, how it is used, and what the returns are on these assets.
Patent assets	List of patents owned by agent, classified by IPC (technology area), validation (country), year left, and legal status (in force, disputed, dropped, etc.). A lot of measures already exist on the asset side.
Contract assets	List of patent rights *in-contracted* by agent, indicating what technology they legally have access and how much they are paying for to use (price).
Contract liabilities	List of patent rights *out-contracted* by agent, indicating what technology is out-licensed of the technology they own. This indicates the distribution of technology in the economic system. The licensing has a "multiplier" effect on the value as more agents can invest in their economic use.
Technology endowment	Know-how in technical areas, not necessarily IP protected. This is knowledge and know-how of certain technical areas. These may be *trade secrets* (an IP) or open access technologies.
Learning	The dynamic process of learning to increase the stock of knowledge needed to have a *constant* development in technology. It is also the integration between science and technology
Knowledge endowment	The human capital formation of employees. This data exists about firms to some extent. A key variable to indicate the dynamism of the agent (and country) is the number of *serial inventors* versus one-off inventors the agent (country) has.
Science & Technology	The "integration" of science and technology by investing in R&D cooperation with research universities and institutes. This is a personal cooperation between individuals making up a team of researchers and inventors. The names of inventors are often listed in patent applications as is prior art literature. Patents are listed in the research literature as the patent application

(continued)

Environment	Characteristics
	must precede any scientific publication. The *investments in research projects* result in patent applications, and the co-publication between researchers and inventor is therefore a measure of learning with purpose of increasing the stock of knowledge that can sustain a constant growth of stock of technology.
Information	Search costs for existing patent, research and research projects, presentation, and so on is a non-negligible cost. Research takes place all over the world, patent ownership is "hidden" from view. The cost of search services is therefore an important factor in reducing the risk (not uncertainty) in involuntary infringement, investment in an area already patented, and so on. The scope of search possibilities is determined by a "*patenting budget*" of inventors/firms.

2.4.2 Institutional Environment

The institutional environment consists of both communication rights and exchange rights. Agents choose messages and the institutions determine allocations via the rules that carry messages into exchange. This is the *social process* that culminates in exchange. In real world situations, these communications rights may be norms or formal rules defined by free trade agreements (FTA) at regional or multilateral levels.

Variable	Characterization
Institutions	Real regional trade agreements or multilateral agreements. This is *the most important independent variable* together with strategy and agent investments.
Regional agreements	These include agreements within economic cooperation areas and are bilateral of poly-lateral.
Multilateral	These include WTO agreements.
Norms	Cultural norms of inventor, innovator, university relations.
Traditions	Traditional knowledge can be characterized as a "communal right." To resolve any private rights, the counterparts may be the community.
Patent systems	The patent system in use by agents. These include international, regional, and national systems. They may be compliant with international agreements such as TRIPS ("minimum standards"), PCT, and EPC (international agreement for "Europe").
IP system	The IP systems in use.
Index	A *patens system index* could be used to characterize the strength of the system. A "*trade index*" could be used to characterize the propensity to trade. In both cases the propensity to patent and trade are given the incentives from the institutions in place. These variables, properly measured and summarized, allow to inform trade policy based in the market efficiency given strategy and institutional arrangements.

Price = Value = *f* (WPT = WTA, environment, [learning], strategy | Institutions). Prices thus differ conditional to the factual institutional arrangements. The listed data captures the dynamic economic system of trade in ideas. Maximizing returns on assets, not profits.

	Economic environment
Contract	Type: License, cross-license, prices, reward medium, resolution process for incompleteness (types)
Agent behavior	
Inventor	Type: Serial or one-off
Resources	HCF
Cost of capital	
Agent ID	Type: Legal entity
Strategy	Type: Business strategy to resolve uncertainty and trade
Investments	
IP assets and liabilities	The IP balance sheet
Patent assets	Licensee's or licensor's IP right
Contract assets	Licensee's IP rights
Contract liabilities	Licensor's obligations with respect to valid IP rights
Technology endowment	
Learning	The stock of knowledge and information
Knowledge endowment	
Science & Technology	
Information	
	Institutional environment
Trade rules	
Regional agreements	Type and by name
Multilateral agreements	Type and by name
Norms	Cultural norms on IP
Traditions	Business traditions on IP
IP system	
Patent systems	
IP system	Type: Systems in use, PCT, EPC, etc.
Index	Patent system strength (laws, enforcement time, cost and access, injunctions, etc.)

2.4.3 *Data to Be Collected*

All the data needed for the value are included in the contract (private information) and public statistics (public information).

2.5 Measures of Outcome and Outcome Behavior

On the basis of data collected, a number of outcomes and outcome behavior can be measured. The goal is to measure how well the countries leverage their human capital formation. Here follows as first selection.

Outcome variable	Characterization
Trade statistics	
Terms of trade	Export prices/import prices by technology area, economic area (regional prices), and HCF area.
Returns on IP by strategy	Explains the sources of terms of trade in terms of trade policy.
Tax-motivated transaction	Transfer prices and transactions not motivated by cooperation but tax-optimization. (This important measure allows to find the motivation of transaction from a trade economic perspective.)
Level playing field	
Trade in ideas index	An index indicating how active and profitable agents (countries) are.
Allocation of resources for invention	Technology areas, amount of investment
Integration of Science & Technology	Number of patents/articles
Agent classification	LDC, developing country, developed country, political north and south
Abusive behavior	
	Policies resulting in not honoring the inventors
Forced technology transfer	Theft
Nonenforcement policy	Theft
Rent seeking	Privateering
Copying	Intentionally using technology owned by others
Financial accounting	
ROA_{IP}	Return on IP assets (RIPA) Firms are assumed to maximize returns on *assets* (income/assets) – not only profits (income) – given the strategy they chose, given the institutional framework.
Price by tech area	
Ownership	Licensing and ownership of technology in use

Part II
Today's Statistical and Financial Accounting Systems and the Additional Data Needed

The previous section outlined the *dynamism* of an economic system centered around trade in ideas, where idea creation is treated as an endogenous process, and a *theory of value* defining which data are needed to measure its efficiency. This section describes *the principles* of today's trade statistics (Chap. 3) – especially the balance of payment – and financial accounting systems (Chap. 4) – especially international standards. A gap analysis is used to identify which additional *data* and which *structural changes* are needed to create a measuring framework for standards and trade in ideas.

Chapter 3
Today's Balance of Payment Statistics and Data Gap

3.1 The Statistical Frameworks and Data Available Today

This section outlines the data that is collected and summarized *today* through national statistics frameworks for *IP embedded* in goods and services, especially current treatment of intellectual property (IP) in the *balance of payment statistics*. Although the focus of this book is on patents (expandable to other IP), this section thus takes the broader IP focus and explains the current treatment of embedded patents.

The next section then identifies *the gap* in *the data*, the *measurement methods*, and the *structure of analysis* of trade in ideas compared with the needs based on the theory outlined in Chap. 2.

This section *reviews and comments on* a report/working paper by A. Maurer and J. Magdeleine from 2019 (or earlier), called "*National Statistics (Statistics Framework) – the treatment of embedded IP*." It deals with the current statistics related to IP *embedded in* goods and services and also proposes ideas forward. The report/working paper was included in the report "Conceptual Framework to Measure Trade in Ideas" of 2019 on which this book expands.[1] The text from the report/working paper is in quotes and indented. Comments and reviews are made to the text in reference to the discussion on trade in ideas (IP).

Information from a number of workshops with representatives of the WTO customs department, the TRIPS department, and other departments of the WTO, and several meetings with a team from the Swedish Bureau of Statistics is also included in the comments and examples, especially in the gap analysis.

---

[1] Government of Sweden, UD/HI, June 30, 2019, Ullberg et al., "Conceptual Framework to Measure Trade in Ideas for Patent Licensing & Transfers (V7.3)," Chapter II. A later version of the report/working paper in Chapter II from the 2019 report was later published in: Magdeleine, J., & Maurer, A. (2022). Measuring International Intellectual Property Transactions in a Globalized World: Current Challenges and Possible Improvements. In A. Taubman & J. Watal (Eds.), *Trade in Knowledge: Intellectual Property, Trade and Development in a Transformed Global Economy* (WTO Internal Only, pp. 171–193). Cambridge: Cambridge University Press. https://doi.org/10.1017/9781108780919.008

E. Ullberg, *Intellectual Property Statistics*, Contributions to Economics,
https://doi.org/10.1007/978-3-031-36386-3_3

3.2 National Statistics: The Treatment of Embedded IP Today

"Intellectual Property (IP) from a business sense is 'Knowledge, creative ideas, or expressions of human mind that have commercial value and are protectable under copyright, patent, [sic], trademark, or trade secret laws from imitation, infringement, and dilution. Intellectual property includes brand names, discoveries, formulas, inventions, knowledge, registered designs, software, and works of artistic, literary, or musical nature. It is one of the most readily tradable properties in the digital marketplace.'"[2]

The current system of statistics focuses on the treatment of embedded (or "embodied") IP. However, this recent business definition ties IP to the "digital marketplace," i.e., software, art, literature, and music.

3.2.1 Current Focus on is on Embedded IP

"IP is an essential component of the *production* of goods and services, particularly for those that are innovative or creative. It is traded internationally:

(i) embedded in the value of traded goods and services, or
(ii) through international licensing transactions, or
(iii) through the outright sales of originals (including offshoring of R&D services)."

However, there is also a major "barter trade" through cross-licensing, which is not captured in these transactions. The broader definition used in today's statics also included R&D services, i.e., there is not a sharp distinction between the IP rights as such, and the goods and services they are embodied in. It is clearly an "extension" of the good, then services, and now goods and services with embedded IP.

"From a statistical sense, The Manual on Statistics of International Trade in Services 2010 (MSITS2010) as well as the revised Balance of Payments Manual 6th edition (BPM6) and accompanying compilation guidance have clarified the concepts of intellectual property in statistical frameworks from previous versions. They present a classification system of related receipts and payments, depending on how IP is transacted internationally focusing on it being *embedded* in products, licensing, outright sales" [and transfers, as rights may be donated for tax and other purposes] "of originals), and at what level of detail (particularly in sub categories of services). However, globalisation, digitalization and tax optimisation have contributed to blur the way transactions are *recorded and compiled* into relevant statistics for policy makers, research and for businesses themselves."

This "blurring" is, in part, because businesses now use different business models where the IP rights are transacted as IP products in their own rights, seeking to maximize the returns.

[2]Online dictionary at http://www.businessdictionary.com/, last accessed on August 31, 2018.

3.2.2 Globalization Requires Broader Institutional Scope

> "MSITS2010 catalogues research and development services as well as transactions in computer software, audio-visual and cultural products. Compilation guides of the BPM6 and MSITS2010, as well as the Guide to Measuring Global Production of the Conference of European Statisticians, which deals with national accounting aspects, provide further guidance as to how compilers could approach some of the challenging measurement issues arising from the globalisation of the economy. But, activities of multinational enterprises (MNEs) make it more complex to value international transactions and assess related transfers of ownership."

To identify third-party transactions, one needs to identify the transactions motivated by tax optimization (cost reduction) and judicial optimalization (legal uncertainty reduction). The business policy related to the management of these assets also matters, especially for MNEs. Some companies have a "decentralized" approach to management of IP assets, whereas others have a "centralized" approach, also resulting in cross-border flows with transfer prices. The remainder is then trade in ideas.

3.3 Concepts Pertaining to IP-Related International Transactions and Current Data Availability

> "Relevant international statistical guidelines have strived to cover innovation and related IP flows in one way or the other, whether in *economic* statistics (System of National Accounts 2008, Balance of Payments Manual 6th edition, Manuals on Statistics of International Trade in Services, OECD Handbook on Deriving Capital Measures of IPPs, OECD Handbook on Economic Globalization Indicators), specialized statistics on *innovation* and *research and development* (Frascati Manual, Oslo Manual, as well as quantitative data on patents, trademarks, etc.)."

Statistics on the *inventions* as non-embedded IP rights represent a conceptual and data collection gap.

3.3.1 The Economic Statistics Related on IP Today

> "*IP products* can be accessed or produced in various ways: within the firm, or through a diffusion process such as accessing open source IP or knowledge, IP acquired from other economic entities (outsourcing of R&D, licensing for the use of IP" [, and the transfer of IP by sale, donating (for tax benefits), etc.], or through the purchase of new innovative products (materials, software etc.)."

Today's *economic* statistics on intellectual property is generally described as taking the following forms:

- *Intellectual property-related products (IPPs), which can take the form of*:
  - *produced assets*:
  - computer software and databases, audio-visual products (music, movies, television, and radio programs),
  - **outcomes** of research & development (R&D),
  - exploration rights (in general no international-related transactions as the client will normally have an entity resident in same economy as owner of rights (in principle, government entity), and
  - *non-produced assets:*
  - **use of** trademark.
- *Exchanges of tangible products*:
  - intellectual property **embedded in goods** transactions, e.g., cultural or creative goods (mass-produced),
  - R&D results **embedded in goods** (e.g., industrial processes), and
  - originals/blueprints exchanged on physical media.
- *Exchanges of intangible products* (*transactions in intangibles*):
  - **Downloaded** products that have a goods equivalent, the difference being the media through which it is delivered/transmitted to clients (i.e., electronic transmission rather than CDs, DVDs, or other physical media).
  - **Licenses for the use of intellectual property**, generally transactions between the owner of originals and producers, distributors that replicate the original, and trademark payments (including franchising type transactions when not possible to separate IP component from other types of transactions, e.g., services).
  - **Transfers** of blueprints, etc. to producers (most probably also in the form of a license) or directly to households such as for 3D printing.

The classification reveals the "struggle" statisticians have to extend traditional "production of goods" to "production" of goods, services, and IP ending up with "tangible," "intangible," and "IP *related*" products. Thus, the IP products are essentially related to, i.e., embedded in, the goods and services in which they are used. In particular, *services* is used to delineate the production and transfers of IP-related goods. One can clearly see the progression of the standards from goods → services → "IP related." However, the operation of the economic system is: ideas → goods → services. The *structure* therefore also needs to be changed.

> "MSITS2010 as well as BPM6 treats the results of research and development (R&D) such as patents, copyrights, and industrial processes as *produced assets* in research and development services (under other business services). Likewise, outright sales of originals are identified separately under the *services that produce* them (e.g. computer software originals are identified as a separate category under computer services). The IP-related product that remains under the *capital account* relates to marketing assets such as brand names, trademarks, logos, etc."

In order to better reflect the idea-based economic system, *all* IP transfers need to be concentrated to the capital account as the proposed view is that they are better described as "production assets," focusing on their *use* process, not only "produced assets," focusing on their *production* process.

> "Further, the breakdown of the item other business services is reorganized, enabling the separate identification of research and development services (R&D). This provides more visibility to an important item in terms of international trade which was previously neglected in international statistical guidelines."

This change seems to be a development in the direction of a separation of IP from R&D services, thus further accommodating economic realities (trying to do that in the current structure).

> "Third, the delineation is clarified between goods, services and different types of licenses for use for some of the most important IP related products in international transactions (i.e. R&D, computer and audio-visual services)."

Again, we see steps in the direction of capturing the most important economic transactions in the current structure, using service. Box 1, seen in Table 3.1, summarizes the current treatment of intellectual property in BPM6 and MSITS2010 guidelines.

3.3.2 The Classification of IP-Related Transactions Today

> "The current classification of transactions relating to the use of originals (copies, reproduction or distribution rights) depend on several factors. Charges for the use of intellectual property in MSITS2010 and BPM6 cover the charges for use of proprietary rights (patents, trademarks, copyrights, industrial processes and designs, franchises, etc.) whether the rights result from research and development, marketing, advertising or other creative or cultural activities (computer software, music, films, books etc.). This item corresponds to the item *royalties and licence* fees in BPM5."

It therefore appears that BPM5 was closer to the economic reality in the aspect of being explicit about the revenues.

> "Charges for the use of intellectual property includes licensing receipts or payment for reproduction and/or distribution of computer software, audio-visual products or other products (i.e. the transactions between the owners of the original/rights and producers who either reproduce the products, or embed them in other products).
>
> However, it is important to note that the item as defined in BPM6 and MSITS2010 (i.e. charges for the use of intellectual property not included elsewhere) does *not* cover the value of end-user licences/copies themselves, which will be covered in the value of goods and services sold ("embedded")."

This change appears to be another attempt to separate the IP value from the services or software value as a whole.

> "The relevant items in the most detailed breakdown provided by MSITS2010 are shown in the Table 3.2 together with the number of countries reporting each individual item for receipts and payments."

Table 3.1 Treatment of intellectual property in BPM6 and MSITS2010 guidelines

Box 1: Treatment of intellectual property in balance of payments statistics			
	Use of intellectual property		Sale or purchase of intellectual property ownership rights[a]
Franchises and trademarks	*Charges for the use of intellectual property n.i.e.*		Balance-of-payments capital account
Outcomes of research and development	*Charges for the use of intellectual property n.i.e.*		Research and development services
Computer software products; audio-visual and related products	**License to use excluding reproduction and distribution[b]**	**License to reproduce and/or distribute[c]**	
(a) Customized products of all types	Relevant service item[d]		
(b) Non-customized products, downloaded or otherwise electronically delivered	Relevant service item[d]	*Charges for the use of intellectual property n.i.e.*	Relevant service item[d]
(c) Non-customized products provided on physical media with periodic license fee	Relevant service item[d]		
(d) Non-customized products provided on physical media with right to perpetual use	Goods		

Source: Manual on Statistics of International Trade in Services, 2010 edition

[a]Covers the case where there is a change of economic ownership of the whole of the intellectual property right in question and the seller no longer has any rights or obligations associated with the intellectual property. This case also includes second or subsequent outright sales of intellectual property rights

[b]Covers the case where a specific product is supplied with the right to use the intellectual property embodied in it, but not to copy it for further distribution. The transactions should be classified under the appropriate goods and services items

[c]Covers the case where authority to reproduce and/or distribute the intellectual property is delegated by its owner

[d]The relevant item is classified under either computer services or audio-visual and related products, depending on the nature of the content provided (see also paras. 3.216–3.220 of BPM6). For example, the sale or purchase of a copy of a software package that is mass-produced, and is obtained by an individual to load onto a single computer is covered by a license to use that excludes reproduction and distribution; this situation would be recorded in goods and services depending on the product (see examples (b), (c), and (d) under software in Table 2.3 in BPM6). If a manufacturer pays for the right to include the software on the computers that it produces, then the payment would be a license to reproduce and/or distribute (charges for the use of intellectual property provided by the owner of the original)

Table 3.2 Reporting of BPM6/MSIT2020 intellectual property related items, 2015

	Receipts	Payments
8 Charges for the use of intellectual property, n.i.e.	142	161
8.1 Franchises and **trademarks** licensing fees	19	21
8.2 Licenses for the use of outcomes of research and development	18	17
8.3 Licenses to reproduce and/or distribute computer **software**	16	17
8.4 Licenses to reproduce and/or distribute audio-visual and related **products**	15	15
8.4.1 Licenses to reproduce and or distribute audio-visual **products**	7	8
8.4.2 Licenses to reproduce and/or distribute other **products**	6	7
9 Telecommunications, computer, and information **services**	180	181
9.2 **Computer services**	116	134
9.2.1 Computer software	10	10
of which: 9.2.1.a Software originals	7	7
10 Other business services	177	181
10.1 Research and development **services**	100	106
10.1.1 Work undertaken on a systematic basis to increase the **stock of knowledge**	26	25
10.1.1.1 Provision of customized and non-customized R&D **services**	25	24
10.1.1.2 **Sale** of **proprietary rights** arising from R&D	25	25
10.1.1.2.1 Patents	8	8
10.1.1.2.2 Copyrights arising from R&D	7	7
10.1.1.2.3 Industrial processes and designs	7	8
10.1.1.2.4 Other	6	7

The most interesting fact is that only 18 countries report patent licensing revenue/receipt and about eight countries report patent transfers. This indicates that the trade in ideas is still rather concentrated to a handful of countries.

> "Although the number of those reporting has not drastically changed for the main item of interest here, that is charges for the use of intellectual property n.i.e. (still in the range of 140 for receipts and 160 for payments), those reporting other main items has increased substantially. Research and development is the item that presents the highest increase, with 50% more reporters (up to approximately 100). The second-best performer is computer services now presenting over 110 reporters for receipts and 130 for payments, followed by audio-visual (around 90)."

One can interpret this development as a changing economic system toward ideas and services.

> "Although MSITS2010 refers to the need to obtain further details, only a very small number of reporters (mainly OECD or EU countries) have compiled information for those. Only 15–20 economies report details for charges for the use of intellectual property, 25 report details for research and development services (of which many are shown as having a 0 value!), and less than 10 economies report data on outright sales of originals or proprietary rights. Finally, data broken down by counterparts in the transactions is available in 40–50 economies (mainly OECD and EU countries). Consequently, it is important that users of the data and compilers further discuss the data gaps and identify ways to better respond to information needs."

The cost and efficiency of the data collection mechanism is therefore a key element needed to be further developed for the proposed framework. Standardized surveys and the use of internet and mobile technology may be a way forward.

> "Given the current structure of IP related transactions the distinction between the different categories from a reporter's perspective may not always be clear (for example software versus research and development, or outright sales of IP versus licensing for use of IP). The suggested granularity of the breakdown of the R&D item is also often seen as challenging by compilers. When it comes to collecting the information, there are basically two main ways of gathering information: reporting by *commercial banks* of international transactions made by their clients (International Transactions Reporting System) and use of *surveys*. The latter offers more options to collect detailed [information]. Given the level of detail sought and the complexity of the subject, one needs to clearly draft appropriate guidance for respondents, which may be particularly difficult when an international transaction reporting system is used (and probably also for general trade in services surveys).
>
> But, the refined classification as presented above should provide a solid point of departure for compilers. Interestingly, the US is spearheading developments on implementing these concepts. It has issued an *enterprise survey* collecting detailed information on international transactions relating to intellectual property."[3]

The proposed framework proposes to make use of these advances as the US development in particular how the third-party transactions are explicitly separated by the surveys. The additional information needed is then the complete hierarchy of the firm's cross-border. This problem has attracted attention for other reasons as well and there are initiatives to create such official registries. However, the rules for "control of ownership" are also used to "hide" ownership relations, including "patent shell companies" in the IP world. This creates uncertainty for the competition playing "Strategy 1" and thus a deterrence mechanism but not good for statistics (transparency). See Chap. 2.

> "Although flows between enterprises (whether affiliated or unaffiliated) should be accounted for in *financial statements*, they may therefore not always be identifiable as specific IP related transactions."

One key issue here is transfer-prices, which are used to transfer payments for the use of IP from "IP holding companies" in tax and judicially preferred countries. Without further separating out these transfer payments, the value of IP transactions motivated by trade, i.e., trade in ideas, is very difficult to compile and incomplete.

> "As for the technological means of delivery, it is important to note that trade involving intellectual property increasingly takes place online through *e-commerce*. The flow of IP protected information is an item of increasing concern to service suppliers around the world, and, in particular, in relation to flows taking place through the Internet or other types of data networks."

The current structure therefore needs to be restructured, not simply refined or differentiated, to capture the value flows. This will better reflect the actual operating of the idea-based economic system.

---

[3] For transaction types covered, see USBEA BE-120, General Instructions, p. 24.

3.4 Measurement Challenges in GVCs

An important issue in international economic law is national sovereignty of states and the role and influence of multinational enterprises (MNEs). The extent of the influence of MNEs on the trade flows of nations is therefore a key policy question. Who are the biggest traders? Today, the MNEs are the biggest privately controlled patent holders and patent traders (and goods and services traders). How much of the IP trade do they represent and who are the others? Which business constellations produce and trade most technology and which university constellations produce and trade most knowledge? Which industry–university or private–public partnerships produce most useful technology?

To answer this, a complete understanding of the hierarchy of these cross-border owned firms is needed. This is not only an issue for MNEs as even the smallest inventors use holding companies for tax optimization or legally stable jurisdiction. In both cases, wealth accumulation from IP licensing often involves transfer prices.

3.4.1 Recording in Practice

> "The establishment of global value chains, global research cooperation and global inventor cooperation is a driving force for the increase of transactions in intellectual property. International competition forces firms to increase their efficiency and develop new products. This phenomenon poses additional significant measurement challenges.
>
> Outside MNEs, IPP-related transfers or payments are tracked using market transactions and the identification of IPP ownership is usually less problematic. Yet, the analysis of IPP use in production still typically requires a complete picture of the global production chain (e.g. "factory-less goods producers", "virtual manufacturing")."

The measurement of simple cross-border transactions is thus more straightforward as long as it involves money, and not barter trade.

> "However, within MNEs it is more complicated. Firstly, intra-firm transactions (i.e. within multinationals) may be affected by (i) the use of transfer pricing and/or more general tax planning issues which distorts the valuation of transactions, and (ii) decisions where to establish multinationals affiliates or headquarters. Transfer prices may significantly differ from the actual "real"-market transaction values. Also, the distinction between flows recorded as trade in services and flows recorded as property income may not always be clear cut. For example, R&D-related transactions or transactions relating to the use of an asset between affiliated enterprises are not always observable. They may also be channeled through convoluted chains of affiliates. Often, payments that are not explicitly recorded, but that are implicitly related to R&D or the use of an underlying asset are instead recorded as property income. In principle, an imputed payment should be recorded (either as a payment for R&D services or charges for the use of intellectual property n.i.e.). As such, more often than not, it is unlikely that such imputations will be made in practice. Nevertheless, if compilers do make such imputations, they should ensure that counterpart transactions (in particular international transactions) should be coherent, requiring *coordination* with other statistics agencies."

The current data may thus be poorly coded, several national agencies may need to get involved (central banks, statistics agencies, . . .), and international cooperation (agreements) between trading nations appears needed to create a "whole."

> "Secondly, defining the location of the economic ownership of intellectual property assets may be an additional barrier to the appropriate measurement of (international) transactions, with the fragmentation of activities, rendering transactions hardly identifiable.[4] Indeed, multinationals may choose to register their patents, in one country rather than another based on "benefit maximizing" considerations."

Companies may also use shell-companies to "hide" ownership from competition, maximizing uncertainty of each other's real IP ownership.

> "The country of registration will be where the *legal* ownership is established. But this does not mean that the country of registration is necessarily the same as the one of the *economic* owner of the IP (i.e. the entity that draws benefits and may encounter losses from the ownership of the asset, whether the producer of the original or the entity providing or using the services or copies of the original). Not all countries may be able to clearly identify these aspects."

The patent administrations rarely update such ownership registries and rarely use the same corporate code as financial registries, and so on. Private services exist for such cleaned-up data, priced for commercial actors. The IP system is clearly not well integrated neither in theory nor practice of trade and statistics. Major advances are needed here.

The recording of data on trade in ideas therefore has major practical "data quality" issues today.

3.4.2 Lack of Practical Recommendations for Compilation

> "As shown above, although recent international statistical standards clarified the classification of transactions, they failed to provide clear recommendations as to *how to compile the information*. When transactions are not observed directly, ownership and the recording of intra-group IPP transfers become uncertain and can only be based on certain conventions. The Guide on Measuring Global Production provides some guidelines in relation to global

[4] In national accounts and the balance of payments statistics, the recording of product transactions is done based on economic ownership change, which is *a fundamental principle*. Economic ownership is defined in paragraph 3.26 of the SNA as follows: "The economic owner of entities such as goods and services, natural resources, financial assets and liabilities is the institutional unit entitled to claim the benefits associated with the use of the entity in question in the course of an economic activity by virtue of accepting the associated risks." As indicated in the *Guide to Measuring Global Production*, accepting associated risks involves the owner's responsibility for maintenance of the asset, which can entail paying for fees to maintain patents, copyrights, or other registrations of the IPP in question. The benefits can usually be derived from embedding IPP in its own production and therefore the subsequent sales of goods or services, or alternatively by licensing other parties for the right to use the IPP in their production processes.

production arrangements, from the perspective of national accounts, as well from the perspective of the trade in services and balance of payments statistics. This complements the compilation guidance drafted by the UN expert group on compilation of trade in services statistics."

The practical aspects range from accounting to compilation.

3.4.3 *Residual Ownership and Standardized Methods*

A key issue in the valuation is to determine *residual* ownership of rights when rights are challenged. However, the first stage is to determine the ownership. These need to take into account the hierarchy of company ownership.

> "The guide on Measuring Global Production provides a decision tree to determine the economic ownership of IPP and related IPP transactions. The starting point is the observation of IPP output or IPP ownership at the level of a certain unit. The obtained information is examined in 4 different steps:
>
> (a) Control ownership of the unit: is the unit member of an MNE (yes/no)?
> (b) Is the unit producer of the IPP (yes/no)?
> (c) What is the main kind of activity (in terms of ISIC) of the unit, or, is the unit expected to use the IPP in its production process (yes/no)?
> (d) Does the unit receive income related to IPPs, or, does the unit pay for the use of IPPs (royalties and licences) (yes/no)?
>
> Of course, to be able to use such a decision tree, appropriate source data are needed. Nevertheless, compilers are strongly advised to implement this decision tree, as this should enable the compilation of relevant and internationally consistent data.[5] Following the establishment of these more detailed guidelines some *specially targeted enterprise surveys* should help improve the data situation, when it comes to the more detailed information sought on IP, intra-firm transactions and location of ownership of IP (e.g. trade in service, FATS surveys, surveys used for the collection of R&D data)."

To clarify the residual ownership of IP and contracted IP rights, information on the judicial system needs to be taken into account.

3.4.4 *Special Purpose Entities and Proposals*

> "More research is needed on how to compile IPP related transactions when SPEs are involved. A clear and separate identification of the IPP income flows received by IPP holding SPEs should be classified under a separate heading such as "IPP related services provided by SPEs" as the provision of these services is not really reflecting economic

[5] Given the difficulties in compiling the information, it is not recommended to reroute the ownership of IP by SPEs.

transactions. Such a practice will help to support a correct interpretation of economic statistics where these entities are located, as well as help improve the data in other countries, whether the related MNEs are established there (parent companies or affiliates), as well for countries where companies have unrelated/arm's length transactions."

The proposals are thus to further refine the data collection based on a *services* approach.

"However, many countries/compilers will not be able to collect accurately such (detailed) information. Also, the increase in the response burden should be acknowledged. A solution for these could be to complement their more general data collection systems with the information collected and disseminated by countries that will be engaging in a more detailed and sophisticated data collection system (most probably the countries which have interest in this information, that is the ones where research and development, and more broadly intellectual property, is significant and where multinationals are involved). However, this can only function if there is an **efficient co-operation between compilers of different countries**. In addition, it will be necessary to have detailed bilateral information published by countries that will be engaging in a detailed data collection system."

The solutions to global value chains framing of the analysis is to propose an institutional cooperation that matches the MNEs. However, such solutions may not be practically feasible. The reason for this complexity may in fact be the lack of appropriate *structure* of the data in the first place. The current approach of services as an entity on which IP-related values can be attached needs to be replaced by *an IP-based structure*. Then there would be three independent trade flows to collect data on IP, goods, and services.

3.5 Characteristics of IP Traders as "Merchanting of Services"

"Finally, an alternative global production arrangement, hence trading, is receiving increased attention in the statistical community. Although not a new channel, global producers and distributors are making a rapid and increasing use of the opportunities offered by **e-commerce**. Consequently, one has witnessed in recent years the development of **new business models when it comes to transactions in intangibles**. This is described in international statistical guidelines as "**merchanting of services**" (see MSITS2010 and UNECE Guide to Measuring Global Production). It can be described as follows: an **intermediary** located in one country arranges the supply of intellectual property, between suppliers and consumers located in other countries, without being engaged in operational activities. This is expected to be applied by MNEs, but is not limited to those. At the time of writing only an attempt of the statistical conceptualization exists, but it seems that this way of doing business is of importance for several major **service trading economies**. Gathering further information on these types of transactions may also **shed more light on how IP is traded**."

Again, we see the extension of the services model to IP but now through new information and communications technology in the form of e-commence. The hope is that by doing so how IP is traded will be revealed. These kinds of intermediary traders are important in making markets work. However, the business is then done in a "network" type of arrangement, not only a multinational hierarchy of control but also a control of flows through contracts and cooperation.

3.5.1 Network Cooperation and e-Commerce

"How can we understand better the characteristics of those engaged in such transactions? How can we analyze the transactions with respect to the size of enterprises, their activities and the relationships between supplying firms and recipients/clients? These are important (trade) policy questions that policy-makers also have. They would need this type of information to improve their decision-making when it comes to determining export strategies as well as for trade negotiations.

Trade by enterprise characteristics (TEC), compiled to complement the traditional merchandise and services trade statistics, can eventually respond to the questions raised in the previous paragraph. Data on international merchandise trade and trade in services are presented according to enterprise size class (measured in number of employees) and type of ownership (domestic whether with only local activities or multinationals and foreign-controlled affiliates). Statistics on Services Trade by Enterprise Characteristics (STEC) have received greater attention in recent years. Eurostat and the OECD recently released a compiler's guide to produce such data. One important feature of such a dataset is that it does not require adding response burden on reporters. However, one needs to have the necessary legal framework and data collection system in place enabling **data linking between trade in services micro data and enterprise characteristics (in principle need for a business register which is used for surveying)**. Keeping in mind **confidentiality constraints**, it would be useful to break down the information to a level of interest to analysts and policy-makers."

Clearly, this approach of services needs to be replaced by another structure to characterize independently, "IP Trade Enterprise Characteristics" (IPTEC). This will better reflect the structure of the economic system. Here, the confidentiality constraints are especially important as patents/IP are unique by definition. A summary statistic may be needed.

3.6 Conclusions

"The level of detail recommended for the type of economic activity is aggregated and it will probably be difficult to suggest more detailed activities given the infancy of this new way of presenting trade statistics. Nevertheless, the current recommendations do suggest a level of disaggregation of services products that would be extremely useful in our context, **separately identifying charges for the use of intellectual property n.i.e. and research and development services.** Although these two items are of high relevance, other recommended items may also be useful.

Finally, presenting the information by **type of ownership** would also be extremely useful to understand the role of multinationals in knowledge diffusion. This can either be analysed in the context of TEC and STEC, as well as looking more closely to foreign direct investment statistic and FATS.

For FDI statistics, it would be interesting to analyse **foreign investment in knowledge intensive activities** as well as obtain a better understanding of the role of **SPEs** in that context. In the absence of statistics applying the **economic ownership principle**, it would be useful to gather more detailed information with respect to SPEs that are used to register assets and act as the legal owner of IPPs, therefore obtaining the revenues of IPP copies or licences to use or reproduce.

> Other sources that should be examined to obtain a better understanding of MNE operations with respect to intellectual property products are operational data portraying the activities of affiliates (sales, exports and imports, employment, assets) as captured by FATS. The information is presented according to the main sector of activity of affiliates, and enables the **analysis of the performance of domestically and foreign controlled enterprises**. This can be particularly interesting to obtain information pertaining to the eventual **ownership of intellectual property assets** and **transactions in intellectual property products such as sales of products with embedded IP, licensing agreements etc., whether domestic or international** (**therefore linking to the balance of payments data**)."

The summary of the current proposed steps appears to be a further extension of the services line of thought. However, since most goods and services contain IP, this necessarily creates a further *differentiation* of the "IP related" trade. Like an "IP cloud" of differentiated items around the traded goods and services. Such differentiation would lead to a "dilution" of perspective on IP and almost impossible to make sense of the economic contribution of trade in ideas (IP) as such. Furthermore, if the data are captured in a good–service–IP-related structure, it will be "impossible" to capture the idea–good–service dynamics of the idea-based economic system.

The proposal in this book is instead creating a new, separate entity for patents/IP trade under which all IP transactions would be gathered in their own rights. The ownership would be done in a similar way, with all IP assets under one heading.

This requires a restructuring of the current statistics resulting in three entities: one for ideas (IP), one for goods, and one for services.

We will now take a closer look at the data gap within the current structure (expanding on services) and a high-level principled approach to a new structure capable of capturing the dynamics of the economic system with an endogenous idea production and trade.

3.7 The Gap in Data to Measure a More Dynamic Economic System

In the previous sections the existing statistics framework outlined the current data collected and possible further divisions in the reporting of intellectual property-related issues. In this section the data collected and possible refinements are compared with the needs for data proposed by the theory of value presented in Chap. 2 to close the gap to measure the value of trade flows from ideas (IP).

3.7.1 The Data Gap in Statistics

The data gap between the current framework and the proposed theory can now be identified. This new data is needed to be collected to compile measurements of the *terms of trade*, *efficiency* of the global markets in ideas, and other micro and macro variables relevant to better inform a policy, research, and business use.

The current reporting of data by countries is also noted to be poor in the field of IP, suggesting that new methods and perhaps cooperation with firms and IP-active industrial organizations are needed to create a "demand" for this data. The Licensing Executive Society (LES) tried this in the Pharma Area beginning in 2004 with some success (Razgaitis, 2004). Trust in the use and sharing of private data remains a top priority.

The data currently available, additional data discussed by the WTO, and the data required according to the proposed theory are now summarized in table form for a data gap analysis (see Table 3.3). See Annex I of 2010 Extended Balance of Payments Services Classification (EBOPS 2010) for current subsections of 10.1.1.2 Sale of proprietary rights arising from research and development. See Table 3.2. A different extension classification is proposed here as an outline. The sale of patents (IP) should have an own category all together or, initially, be under 10.4 or 10.1.2 related to the "Work undertaken on a systematic basis to increase the stock of technology" (to distinguish this work from 10.1.1, work to increase the "stock of knowledge"). This is important as IP rights are not granted for "knowledge" (discoveries of nature) – it belongs to everyone – but for "ideas" which are creative arts by humans and quintisentially private in nature.

3.7.2 A Change in Structure Needed

Although data can be added to the current framework' structure and division of transactions to compensate for the needs of the new economic system of trade in ideas, goods, and services, the structure itself is less suitable for trade policy in trade in ideas or "things in action" as discussed above. Today's "fetish" approach to value appears rooted in the trade in goods or "things in possession." (See above the discussion on chose in possession (goods) vs. chose in action (IP).) Even with an expansion on services, as discussed in the previous section, there will be no simple way to capture the dynamics. In order to better capture the dynamics of an idea-based economic system (idea–goods–services system), we need a structure that is based on trade in "chose in actions" that replaces the current approach of "chose in possession." A new top-level "idea classification" is needed before and at the level of goods and services classifications.

Human capital is formed through research often directed by demands for new knowledge needed to solve particular technical problems (some are serendipity findings through). A differentiation is needed between "stock of technology" and "stock of knowledge" where the patent (IP) protected technology has its own differentiation. Some state-of-the-art solutions are then protected using intellectual property rights, creating tradable assets, which are *allocated* trough world *trade* in ideas (but also nonmarket mechanisms and theft). The allocation of IP assets is

Table 3.3 IP statistics

Transactions (additional disaggregation and classification based on proposed theory in italics)	Chose in possession focus		Chose in action focus
BPM6	Current	Discussed	Theory (IP centric)
ROYALTIES AND LICENSE FEES			
8 **Charges** for the use of intellectual property, n.i.e, proprietary rights not included elsewhere (BPM5 = Royalties and license fees)	x		x
8.1 Franchises and trademarks **licensing fees**	x		x
8.1.1 Franchises licensing fees			x
8.1.2 Trademark licensing fees			x
8.2 Licenses for the use of outcomes of research& development	x		x
8.2.1 Licenses for the use of IP by type			x
8.2.1.1-8 Patents, Trade secrets, . . ., future IP			x
8.2.2 Licenses for the use of other proprietary rights of R&D by type			x
8.2.2.1-4 Licenses for the use of exploration rights of research, . . .			x
8.3 Licenses to **reproduce** and/or **distribute** computer **software**	x		x
8.4 Licenses to reproduce and/or distribute **audio-visual** and related products	x		x
8.4.1 Licenses to reproduce and or distribute **audio-visual products**	x		x
8.4.2 Licenses to reproduce and/or distribute **other products**	x		x
SERVICES INCL. SALE OF PROPRIETARY RIGHTS			
9 Telecommunications, computer, and information **services**	x		x
9.2 Computer **services**	x		x
9.2.1 Computer software	x		x
of which: 9.2.1.a **Software** originals	x		x
10 Other business **services**	x		x
10.1 Research and development **services**	x		x
10.1.1 Work undertaken on a systematic basis to increase the stock of knowledge	x		x
10.1.1.1 Provision of customized and non-customized R&D services	x		x
10.1.1.2 **Sale** of proprietary rights arising from R&D	x		x
10.1.1.2.1 Sale of IP rights	"different sub-structure in use"		x
10.1.1.2.1.1-8 Patents, trade secrets, . . .			x
10.1.1.2.2 Sale of other proprietary rights			x

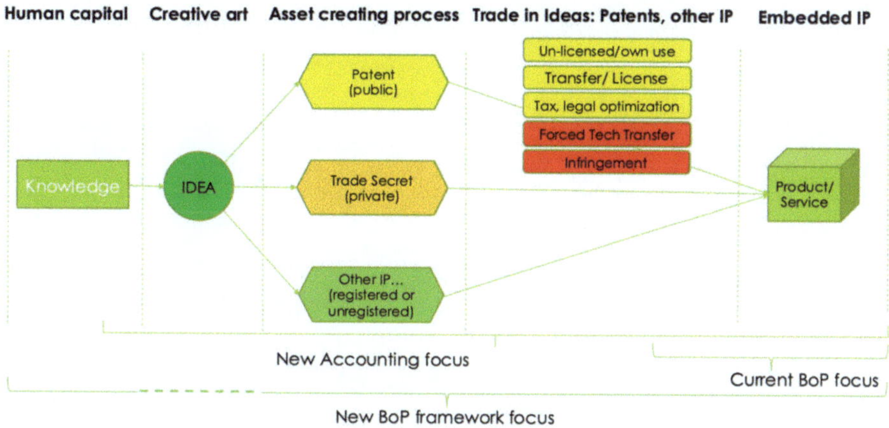

Fig. 3.1 New focus on BoP to include more of the dynamics of the trade in ideas (IP)

ultimately motivated by the demands for more competitive goods and service for the world export markets (see Fig. 3.1).

This shift in the statistics framework principles better captures the *dynamics* of the economic system by including the *time dimension*, as trade in ideas comes first, followed by, often years later, trade in goods and then services that use the new ideas. It also better captures the structure of the economic system, where markets in IP play an important role in coordinating research and development activities. (See the discussion above on the theory of value.) The focus is then not only on the "output" (such as GDP or FDI) or the factor "input" (such as human capital) but also on critical elements of the *mechanism* of trade in ideas, giving the incentives to arrive at an economic, socially, and environmentally preferable output. These *incentive-compatible rules* normally need to be discovered through experimentation.

Improved financial accounting standards are needed to capture the private information on patent transfers, licensing fees, and transaction of other rights, as well as IP ownership, corporate ownership, the operating of new business models, and the operating of the economic system as a whole. This is the topic of the next chapter.

Chapter 4
Today's Financial Accounting Standards and Data Gap

The new business models of early internationalization, networking, and integration with universities and other IP-centric models are characterized by cooperation between entities of highly specialized knowledge and inventiveness and increase the demand for more *differentiated information on the operation of firms*. In these twenty-first-century business models, it is necessary to effectively manage the *inventiveness* of businesses, including *investment* decisions in the area of IP, which requires new financial information on these business processes. Such information, in turn, provides for better information on the operation of the new economic system as a whole to businesses, policymakers, and universities.

Today's financial accounting standards were developed in the early twentieth century partly (or even mainly) in a response to the separation of responsibilities between the *investors* (owners) and hired *management*. Owners could then get the necessary information on the *performance by management* in executing the desired strategy decided by the owners.[1] Today, when creating and adopting new technology is increasingly a cooperative activity among firms specializing in inventing and implementing, a further differentiation is needed to measure the performance of management in creating returns from both purely operational assets and creative assets. According to some estimates, 85% of the value of firms come from intangible assets (IA), of which IP is a large part, and only 15% from physical assets. This means that the current financial accounts only explain bout 15% of the value of firms (see Lev, 2018). A USPTO study estimated that 40% of US traded goods have IP embedded in them. This is likely >50% today. This critical division in management responsibility will be further elaborated on in the proposed framework chapter.

The focus of this book is on the management of the patent (IP) rights that firms can own and trade in their own rights. Access to these intellectual property-based *production factors* need to be more clearly visible (transparent) for investors and general management just as physical assets are.

[1] See, for example, "An introduction to corporate accounting standards" (Paton & Littleton, 1940).

E. Ullberg, *Intellectual Property Statistics*, Contributions to Economics,
https://doi.org/10.1007/978-3-031-36386-3_4

Today, when human capital formation is much higher in general in the society and provide the core of competitive advantage (and creates comparable advantage for countries), the creative side of highly educated people must be better integrated into the firms' activities. Not only does this make sense but educated people *want* to use their knowledge, to solve problems, and to learn more; thus, they choose firms where this is possible or create more and more own start-ups where such learning opportunities are inherent to the organization. Stock options are since long more important than initial salary in many cases in these companies. This requires a better understanding of managing the inventiveness of people working in firms, the creative "input," rather than just the manufacturing or service "output" from "off-the-shelf" input.

The process of inventing and designing is difficult to scale as it is a *small team* operation (or several parallel teams), which is in stark contrast to the readily scalable process of manufacturing/performing of services.

For these and other reasons, the universities have now increasingly taken on an *economic* role in society where young students and senior professors and researchers try out new ideas (see Ullberg, 2015b).

The future management challenge includes separating out the *management of inventive activities* that deliver the new technologies and methods needed for greater productivity – both in the production of technology itself and in the production of goods and provision of services using the technology – in companies and the economic system as a whole.

4.1 A New Challenge for Management

From a management perspective, the question is then how we make sure each *individual* is a talent to the organization. How do we think about that? We need processes and rules to create continuity, but it is *empathy* that is required to motivate *people* to invent.

It is this management process – or its outcome – that has to be measured in a standardized way, partly through an upgraded financial accounting system. The performance of management of inventions can then be better evaluated. The institutional performance of trade in ideas can then also be measured, given the communications and exchange rights in the market in patents – the "trade rules" discussed earlier – used to coordinate activities from specialized inventive firms, and, when summarized, creating statistics on trade in ideas (see Fig. 4.1).

To this end, the existing international financial accounting frameworks need to be amended and even partly restructured in order to reflect the management efficiency of this new management area, thereby providing the data needed for the statistics frameworks for the economic system as a whole. Visibility of inventions need to be differentiated. Such data will allow us to compare *economic systems* relying on different trade rules, processes, and management culture and skills.

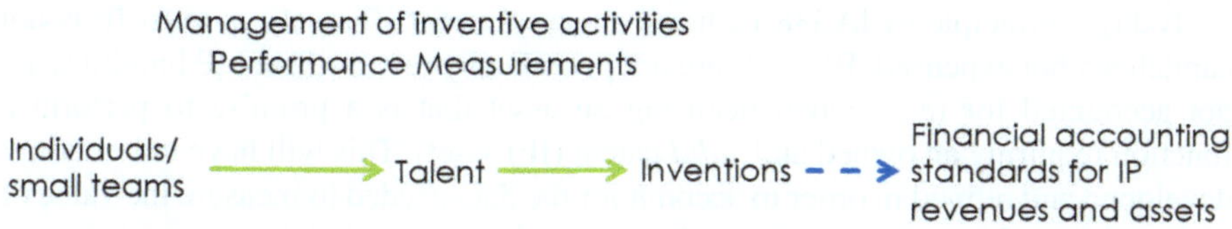

Fig. 4.1 Empathic management of talented people needed to create competitive inventions for a market in ideas whose outcome, including specialization in certain technologies and financial Returns on IP Assets (RIPA), can then be captured in the financial accounts

Before presenting the new framework proposal (Chap. 5), the current standardized financial accounting framework is briefly described in light of these management information needs, and then the gap between it and the proposed theory of value and basic needs for management of inventions is outlined, as was done for the statistical standards (Chap. 3).

To make this description understandable and meaningful to a broader public outside the standards setting community, only the treatment of IP and IP transfers and licensing are included in the discussion together with the investment in R&D projects generating the IP assets. These are the key elements to measure in order to create statistics on trade flows from IP and IP investments. They also capture the investment strategies firms use to reduce uncertainty and risk.

A summary table is included at the end of each section relating the current standards provisions to the *income statement and the balance sheet of firms*. The data for statistical purposes is then drawn from these accounting statements in annual reports and specialized surveys.

4.2 The Accounting Frameworks and Data Available Today

The current accounting frameworks for the recognition of revenues, the International Financial Reporting Standard IFRS15, and the International Accounting Standard IAS38 of the International Accounting Standards Board (IASB) focus on the accounting valuation of *trade flows* and are based on the *contract* between parties in a transaction. It replaced previous IFRS and US GAAP standards, creating a single model for revenue recognition from customers.

The objective of IFRS15 is "… to establish the *principles* that an entity shall apply to report *useful information to users* of financial statements about the nature, amount, timing and uncertainty of revenue and cash flows arising from a contract with a customer."

This principle of IFRS15 – that all transactions emanate from the contract – is suitable for the expansion of goods and services products to "idea (IP) products." It also allows to make a direct connection between the contract and the underlying patent (IP) assets.

Today's principle of IAS38 is, however, weaker for IP assets as most IP is not capitalized but expensed, IP assets are not periodically revalued, and IP liabilities are not accounted for (e.g., when licensing an asset that is a promise to perform a function requiring an owned and *valid* patent (IP) asset). This will have to be further developed and refined in order to account for the data needed to measure the value of trade in ideas, the terms of trade, and the gains from specialization.

There are at least two key aspects of this "dual" approach that are important for the purpose of trade in ideas:

(a) *the principles for the recognition of revenues*, which is based on the terms of the contract, and
(b) *the criteria for recognition and measuring the intangible assets* of which IP is a major part and patents the most desirable, the principle being that the (separable) referenced assets in the contract must be *disclosed* (whether they are other patents; other IP such as trade secrets, data; money, including cyber cash; equity; or even work).

The dual approach thus relates directly to the *income statement* and the *balance sheet* of the firm. The discussion will therefore be made along the lines of these statements.

4.2.1 Income Statement: The Principles for Recognition of Revenues

The current treatment of IP in international financial accounting is analyzed and commented on comparing with the needs given ty the proposed theory. The comparison is summarized in an income statement format at the end of the section. From this microeconomic data, a statistic can be consolidated at a macroeconomic level. The comparison is done at a level of principle, i.e., how the revenues and the costs related to IP are recognized.

4.2.1.1 A Lack of Dynamism

The economic system based in ideas is *highly dynamic*, with a logical sequence of activities of ideas → goods → services as described earlier.

The IFRS15 standard currently focuses on "goods and services" and less explicitly on "ideas" when it comes to what is transacted. It focuses on the *transfer* of the products promised, not the right to use them. This is clearly a "*chose (thing) in possession*" or a "fetish" approach, where a physical "thing" is transferred, and less of a "*chose (thing) in action*" approach, where a right to use – or not to use, i.e., block or "sit on" – is transferred (see Chaps. 1 and 2 on the "legal doctrine" of assets as a "chose in possession" or a "chose in action"). This is applied even for services.

A *right to use* is implicit, and many, if not all, goods are not allowed to be sold if they are not accompanied by the right to use (certain medicines, chemicals, electrical supplies requiring a certificate to use, insurance demands, certain weapons, radioactive material, etc.).

From an economic perspective, an action right approach therefore makes sense not only for ideas (IP) but for most – if not all – goods and services as well. Due to legal and other *constraints* of usage, property rights on "things" can instead be seen as a "bundle" of rights. This approach of a "bundle of use rights" to define "property" *unifies* the understanding of ideas, goods, and services and makes economically meaningful distinctions between them possible as well. This has implications for financial accounting standards and the classification of revenues.

However, currently the standards build on the physical object being transferred, which is then extended to services and licenses of rights. IFRS15 states that:

> "IFRS15 establishes the *principles* that an entity applies when reporting information about the nature, amount, timing and uncertainty of revenue and cash flows from a *contract* with a customer. Applying IFRS 15, an entity recognises revenue to depict *the transfer* of promised *goods or services* to the customer in an amount that reflects the consideration to which the entity expects to be entitled in exchange for those *goods or services*." (italics added)

Revenues are thus recognized as "goods or services" and when they are transferred. There is no *explicit* recognition of intellectual property as a "thing" with its own rights that can be transferred. The legally construed "*chose in action*" approach appears to be neglected.

The view of the proposed theory is that patents (IP) should be seen as *production assets*, thus a "thing" in its own rights, similar to a good or a service but not a physical good or service but a *right* to produce a good or perform a service. You can't eat patents (IP) but you can eat or use manufactured goods and performed services made efficient by the help of these "producer rights." There are no other rights recognized as a "thing" under the apparent "chose in possession" approach in accounting (such as land lease rights, concessions, fishing rights, exploration rights, etc.).

To develop a good or a service, you first need the technology (often including patents) and other IP. This is a *dynamic* question – first ideas, then goods, and then services – that is less *directly* addressed in the current way of thinking in accounting. IP is recognized as revenues (under the goods or services heading) but not under its own kind, i.e., a chose in action. This development process of new technology ought to be seen as being implemented in a *producer market* for transfers and licensing of technology and other IP.

The producer market in patented technology is today a global one, increasingly using joint ventures, in-house projects, creation of jointly owned patent pools, mergers and acquisitions of IP-intense companies, and transfers and licensing to acquire these "producer rights." This process is necessary to be made visible – or at least its outcome – in order to evaluate the quality and efficiency of the *management* of such technology production process. The dynamic question is therefore essentially a *classification of revenue* issue to be able to capture the value flows from the logical steps from ideas to goods to services.

Separating the accounting of revenue streams by "types" in this dynamic process is therefore key to understanding the management of inventions. A proposal for how to make this division into types is developed further below.

To differentiate these "types" into an accounting framework, it seems sufficient to expand on the treatment of "goods and services" flows and restructure the "licensing of IP" to an own type of revenue stream. This is in fact what the leading patent developing and licensing firms in the world are beginning to do to indicate where the revenues come from. Some examples will be given in the text.

4.2.1.2 Current Principles for Treatment of Revenues from Licensing of IP

In the current IFRS15 standard,

> "Goods and services *revenues* are recognized when the *performance obligations* outlined in the contract are *satisfied* in the amount of the transaction price." (italics added)[2]
>
> "A performance obligation is *a promise to transfer* to the customer a good or service (or a bundle of goods or services) that is *distinct* (IFRS 15.22). A good or service is distinct if the customer can: a) benefit from the good or service either *on its own* or together with other resources that are *readily available* to the customer, and b) the good and service is *separately identifiable* from other promises in the contract. Finally, a performance obligation is satisfied by *transferring* a promised good or service to the customer.
>
> For revenues from *licensing of intellectual property*, the main question then is to determine whether such a licensing of IP constitutes a) *a distinct good or service*, and if so, b) whether the related performance obligation is satisfied *over time* or *at a point in time*."

The *principle* of IFRS15 for the recognition of revenues is the "transfer" of a good and service. This principle is applicable for the "transfer" of patent/IP rights both for transfers of the patents/IP and the licensing of right, which "transfer" the right to use.

4.2.1.3 A Distinct License

> "A license is usually *distinct* but sometimes a customer benefits only from a *combined* entity's output and therefore a license forms a part of wider performance obligation. For example, an embedded software in a machine or on-line 'cloud' service cannot benefit the customer without the machine or the service."

This issue is actually a fundamental issue in the software licensing debate where the freedom to access and improve on "embedded" software is taken to a level of a

[2]The text in this section comments on and reviews: (1) selected IFRS community's documents on recognition of revenues as summarized in: https://ifrscommunity.com/knowledge-base/revenue-from-intellectual-property/, and (2) the IFRS15 and IAS38 standards themselves. The texts commented on are referenced in quotes and paragraphs are indented. Italics are added.

principle.[3] The question related to trade in ideas is thus whether the IP is embedded in a good or a service, in which case it is not distinct, or non-embedded, in which case it is distinct.

The value of the embedded IP is thus not explicitly accounted for, whereas the value of the non-embedded IP is. As mentioned earlier, a majority of traded goods and services have IP attached to them. The contribution of the IP to that value traded is thus "embedded" in the accounts for goods and services.

There is a range of contracts used in the IP licensing. In capture-period contracting, new technology invented during the period is included in the contract. If the contract is a "guillotine" contract, the rights cannot be used after the end of the license period. If the contract is a "patent life" contract, the rights can be used for the life of the patent. Many other contracts exist.

> "IFRS distinguishes between two types or license agreements: a) a right to access the IP as it exists *throughout* the license period, or b) a right to use the IP as it exists at *the time* when the license is *granted*. In the first case, the related performance obligation is satisfied *over time*, whereas in the second the performance obligation is satisfied at a point in time at which the licence is granted to the customer. However, 'revenue cannot be recognised before the beginning of the period during which the customer is able to use and benefit from the licence (IFRS 15.B61 and IFRS 15.BC414)'."
>
> "The criteria to decide between the cases is that in the first case (IFRS 15.B58) a) the IP is significantly affected during the license period (e.g. further developed); b) the customer is exposed to these changes and c) these changes do not result in the transfer of a good or services (e.g. software updates that are treated as distinct services). This is to say that the customer is exposed to the benefits or risk of changes and is presumed to use the latest IP."
>
> "These assessment of these criteria cannot be affected by *other* performance obligations in *the contract* (IFRS 15.BC410). For example, when software updates are treated as a distinct service, they are not taken into consideration in determining when and how control of the software is transferred to the customer."

The clear differentiation of the contract's *performance obligations* is thus the basis of separating out the value related to IP licensing. The concept thus permits to differentiate the obligations by technology areas, fields of use, geographical markets, and over time.

4.2.1.4 The Payment Terms Are Separate

> "The *payment terms* are not included in the criteria. This is because the IASB considers payment terms as not indicative of whether the licence provides the customer with a right to access or right to use the intellectual property and therefore they are not indicative of *when* the performance obligation is satisfied (IFRS 15.BC412(d))."

---

[3] This is a fundamental issue of debate in software licensing resulting in the creation of the GNU license. The idea being that users should have the right to access the software and improve it. A little like reading a book by Shakespeare and making a better story without paying for the use of the original material, except in attribution. This goes against the notion of property rights on creative human ideas.

This principle is important as payments for the IP is typically both in other IP, including information, equity, and money, and even work (for inventors). There may be net-payments as well, which are contracted and paid at some point. Even zero-royalty payments fall into this category.

Thus, separating the payment terms from the recognition of revenues allows the payments to be purely a *balance sheet* transaction. On the balance sheet, the clearing of the contract may thus result in a wide range of IP assets and financial assets.

> "Yet another approach is used by US GAAP where IP is distinguished between *functional and symbolic* intellectual property (IFRS 15.BC414M.) Such discussions appear to be the dead-end result of the 'fetish' approach to what is transferred."

4.2.1.5 Not Distinct Licenses

> "If a license is *not* distinct, entities should determine whether the license is a *primary or dominant component* in the performance obligation (IFRS 15.BC407). If so, the specific provision relating to *licensing* of intellectual property should be applied. If the license is not a primary or dominant component in the performance obligation, the *general criteria* for satisfaction of performance obligations apply."

However, IFRS15 does not contain any specific criteria for determining whether a license is a primary or dominant component in the performance obligation.

See Table 4.1 for a summary of revenue recognition. The sale and development of IP are discussed in the next section.

4.2.1.6 IP as Assets Necessary to Reflect Both Risk and Uncertainty

The financial statements should reflect risk and uncertainty. Since the investments in ideas are genuinely uncertain, firms use different investment strategies in these assets to resolve the uncertainty (see Chap. 2). This means that IP needs to be accounted for as investable assets to correctly reflect the *management* of inventive activities. Treating IP as assets is generally not the case today, with few exceptions (Goodwill), and standards differ between US GAAP and IFRS. Separation in terms of risk and uncertainty is needed for a twenty-first-century financial accounting system.

Goods – low risk – often have a legally required warranty, then technical service are added, then finance (access/leasing), then, with internet, controlling of performance at the customer's end, and finally management of risk in the customer's business (Ullberg et al., 2002). These historic "steps" of management of risk motivate the separation of goods and services.

When it comes to ideas, patents, and IP, managing the uncertainty is a sociological problem of trust, and cannot be resolved only by information and transparency (you cannot reveal intent, which may also change). The financial information must therefore cover the strategies firms use to cover the cost of managing uncertainty and risk. These strategies have very different investments in IP assets and IP liabilities, which leads to the necessity of valuing these assets and liabilities on the balance sheet. This will be discussed in the next section.

Table 4.1 Financial accounting to measure the performance of management of inventions

Financial accounting to measure the performance of management of inventions today	
Income statement	Chose in possession focus
Revenues	
Principle: Recognition of revenues comes from contracts on goods and services which have single or multiple performance obligations with customers (often disaggregated by entity's operating/market segment)	IFRS15.1–2
Promised goods and services include the granting of licenses of IP	IFRS15.26(i)
Goods and services	
Contracts where IP is not a primary performance obligation	IFRS15
Licensing of IP	
Risk adjusted licensing revenues from, or treated as a single, performance obligation, sometimes disaggregated by use segment, based on portions of an entity's IP assets	
Contracts may include IP updates over the time of the contract (a "capture period") or for specific IP or cross-licensing with net payments	
Distinct (non-embedded, non-embodied, un-bundled) IP *only*	IFRS15.B52–53
At a point in time (such as a fixed fee or royalty)	
Over time (such as annual fixed fee or royalty based on use)	
Non-distinct (embedded, embodied, bundled) where IP is the *primary* component of the performance obligation	IFRS15.B54
At a point in time (such as a fixed fee or royalty)	
Over time (such as annual fixed fee or royalty based on use)	
Transfer of IP	
Sale and purchase of IP; affects IP assets on balance sheet	IAS38
Exchange of IP	
Exchange of IP assets without any payments; affects IP assets on balance sheet; a net-payment may be part of transaction	IAS38
Cost	
Research phase	
Research phase: Research costs are sometimes depreciated but	IAS 38
Development phase (IP)	
Recognition of cost under a cost model or a revaluation model	IAS 38.72–73
EBITDA	
Depreciation: amortization and any impairment losses	IAS 38.74
Intangible assets are depreciated for their economic life, i.e., 10 years	
Research costs are depreciated over shorter periods, i.e., 3 years	
EBIT	
Financing with IP collaterals	
EBT	
Tax on IP	
EAT	
Income from IP	

4.2.2 Assets and Liabilities: The Criteria for Recognition and Measuring the Intangible Assets

The sale and development of IP are covered in this section. To evaluate the performance of management of inventions the return of IP assets, here referenced as "RIPA," is a key measure. It indicates how good management is at turning the firm's IP assets into earnings. However, to calculate RIPA we need not only the returns, which are calculated from the income statement as outlined in the previous section, but also the *IP assets and IP liabilities* of firms, which are part of the balance sheet. The treatment of IP as assets and liabilities is now discussed under the current standards, and a gap analysis is made in comparison to the proposed theory of value, which allows the calculation of these returns.

In addition, the *cash-flow from assets* must also be calculated, which requires valuing the *investments* in IP assets and liabilities. Such measures allow to *estimate* the net present value of investments under risk. Since uncertainty needs to be resolved prior to trade through specific investment strategies (see Chap. 2), these investments need to be part of the estimate, even if they only indirectly result in cash-flows. The investments are thus a combination of investments under uncertainty (to deal with the trust problem) and investments under risk (to deal with the economic problem). This is a major departure from traditional investment calculations but necessary in the case or uncertain (and risky) IP assets. IP liabilities refer to companies that are out-licensing their technology to others, thus have an obligation toward a third party.

Other measures are also possible when including IP as assets. See discussion under "framework," Chap. 5.

4.2.2.1 The Treatment of IP Assets in Financial Accounting Standards

The treatment of IP assets and liabilities takes place under the headings of *sale and purchase or exchange* of intangible assets, *recognition of cost* of intangible assets, and *internally generated* intangible assets, which are classified into the research phase and the development phase.

The (1) *sale and purchase or exchange* of *intangible assets* are covered under the IAS38 standard, with some exceptions. Any expenditure that does not result in recognition of an *intangible asset* within the scope of other IFRS is within the scope of IAS38. This applies to patents, licenses, and other IP; software; customer relationships, including contracts and databases and other assets.

The (2) *recognition of cost of intangible assets* and (3) *internally generated intangible assets* are also covered by the standard. To facilitate this process, IAS38 classifies the generation of the asset into a *research* phase and a *development* phase (IAS 38.51–52).

4.2.2.2 The Sale and Purchase or Exchange of Intangible Assets

4.2.2.2.1 Intangible Assets

> "*Intangible assets* are non-physical, non-monetary and are *identifiable*. An asset is identifiable if it either is *separable* or arises from *contractual or other rights* (IAS 38.12). An asset is separable if it can be separated or divided from the entity and *sold, transferred, licensed, rented or exchanged, either individually or together with a related contract, identifiable asset or liability*. This includes technology-based intangible assets, i.e. patents (IP) (IFRS 3. IE39-IE44). Assets that do not meet separability criterion or contractual-legal criterion are included in *goodwill*. Exclusions include deferred tax assets and goodwill (IAS 38.2–3). This section is based on IFRS documents."[4]

These definitions leave a lot to be asked for to account for the value of sold, purchased, or developed patents and other IP.

4.2.2.2.2 Embedded IP

> "Some intangible assets are contained in ('embedded' or 'embodied' in) or on a physical substance. For example, computer software can be pre-installed on a computer or can be written on external drive and available for installation on any device. Judgement is needed to assess which element is *more significant* (property, plant and equipment are accounted under IAS 16).
>
> The general rule is that if an intangible asset is not *an integral part* of the related hardware, it should be accounted for separately under IAS 38 (IAS 38.4). Examples include software that can be installed on *any* hardware and documentation for a *patent* or a *prototype*."

Interestingly, the view in the proposed theory is that documentation (know-how), patents, and prototyping (not production) are included in the IP side. Serial production of a prototype then makes it a traded good or a traded service. The value of the embedded IP needs to be accounted for separately, or, if at least when the majority of the value of a good or service comes from IP, as a category of good or service with embedded IP. This will make visible IP, IP intensive goods and services, and then goods and services without IP, i.e., generic technology.

4.2.2.2.3 Intangible Assets or Inventory

Any intangible asset that will not be "consumed" after one use, such as patents (IP), can be treated as an intangible asset within the scope of IAS38 with its amortization presented *below* EBITDA, i.e., it is not treated an inventory (that is consumed), but a

[4] See referenced text on IFRS documents: https://ifrscommunity.com/knowledge-base/intangible-assets-scope-definitions-disclosure/.

"production asset" (that is "used" and amortized). This means that the standard in principle supports the calculation of, for example, "return on IP assets" (RIPA),[5] similarly to "return on assets" (ROA) for physical assets. The standard is thus already quite developed for these typical performance measures, also for IP. (Where it is lacking is in the recognition of value.) More on this in the next chapter.

4.2.2.2.4 Revaluation of IP Assets

IAS38 allows a policy choice when measuring intangible assets: a *cost* or *revaluation* model (IAS 38.72–73). Intangible assets are initially measured at *cost* (IAS 38.24).

> "Under [the] *cost* model, an intangible asset is carried at *cost* less any accumulated amortization and any accumulated impairment losses (IAS 38.74). Under the *revaluation* model, an intangible asset is carried at its *fair value* (i.e. revalued amount) less any accumulated amortisation and any accumulated impairment losses." (italics added)

The cost model is the model most favored by the theory as the value is uncertain. With the cost model, one can estimate the returns on IP and the cash-flows from investments that the management of inventions have achieved.

4.2.2.2.5 Measurement of Acquired Assets and Liabilities

> "The acquirer measures the identifiable assets acquired and the liabilities assumed at their acquisition-date *fair values* (IFRS 3.18–19), with certain exceptions."

The fair value ("market value") principle is a contested approach to valuation ascending in the 1990s to better reflect the current balance sheet value of assets (allowing higher debt). Such accounting may result in a short-term fluctuation of the asset value, affecting the capital base and creating a *real risk* of insolvency based on "*artificial*" *values*.

For trade in ideas purposes, a cost approach ("accounting value") is sufficient to calculate the return on IP and thereby evaluate the *actual performance* of management of inventions. Such management evaluation is the purpose of the accounting extension for trade in ideas.

[5]RIPA is proposed acronym for return on assets (ROA) for IP assets. See chapter on proposed framework.

4.2.2.2.6 Mergers and Acquisitions

> "IAS 38 provides application guidance for separate acquisition of intangible assets (IAS 38.25–32) and acquisition as part of a 'business combination', i.e., mergers and acquisitions (IAS 38.33–37)."

This principle has similarities with embedded IP goods and services. It allows to separate out IP to some extent.

4.2.2.2.7 Exchange of Intangible Assets

> "Exchange of intangible assets mirror the requirements of those of tangible assets (IAS 38.45–47)."

This is an important requirement, as many transactions are barter trade for other IP and assets rather than money.

4.2.2.2.8 Fair Value of Intangible Assets

> "The fair value of an intangible asset is reliably measurable if (a) the variability in the range of reasonable fair value measurements is not significant for that asset or (b) the probabilities of the various estimates within the range can be reasonably assessed and used when measuring fair value. If an entity is able to measure reliably the fair value of either the asset received or the asset given up, then the fair value of the asset given up is used to measure cost unless the fair value of the asset received is more clearly evident." (IAS 38.47)

According to the proposed theory, such future values are *impossible* to estimate as they are genuinely uncertain ex ante any market test or use. Only in narrow use-case where many market tests have been conducted, can such a fair value be valued as risk (known variability). Thus, the cost approach seems to be the most interesting for the reasons stated above.

4.2.2.3 The Recognition of Cost of Acquired Intangible Assets

The cost approach is the preferred method to value IP due to the uncertainty of its future value. Since much of IP are used for strategic purposes to reduce uncertainty and create trust in each other's action not to sue or otherwise harm (see Chap. 2), the future economic benefits need to take into account these "preconditions" for economic value through trade.

An intangible asset is recognized when it meets all of the criteria of *identifiability*, *probability*, and *control* of the future economic benefits and *reliable measurement* of cost (IAS 38.18,21).

The strategic approach to trust appears to be an addition to current standards that focus on *risk* (identifiability, probability, control) of future economic benefits. The criteria of "control" (ownership of rights, contracted rights) has relevance to *uncertain* values as the strategies are based on access to the IP assets. The reliable measurement of cost for both the strategic and the operational IP assets therefore needs to be considered as assets. This section contains references to IFRS documents.[6]

> "An asset is identifiable if it either is separable or arises from contractual or other legal rights. The probability recognition criterion is always considered to be satisfied for separately acquired intangible assets. Human capital formation, such as training, paid by the firm cannot normally be recognised as assets because of insufficient control over the expected future economic benefits.
>
> Separate acquisition of intangible assets is not to be confused with acquisition of services that are used by the firm do develop an intangible asset internally. In such a case, the requirements for internally generated intangible assets apply.
>
> IAS 38 has more stringent requirements concerning capitalisation of subsequent expenditure on intangible assets. Paragraph IAS 38.20 states: '. . .In addition, it is often difficult to attribute subsequent expenditure directly to a particular intangible asset rather than to the business as a whole. ... This is because such expenditure cannot be distinguished from expenditure to develop the business as a whole.'
>
> The cost of an asset acquired as a part of a business combination, i.g., a merger and acquisition transaction, is its fair value at the acquisition date.
>
> The probability recognition criterion is always considered to be satisfied for separately acquired intangible assets."

One would consider the criteria for intangible assets a little too vague for IP assets. A more stringent approach is needed.

Consideration of enforceability seems important as the courts vary a lot in quality. For example, firms rarely validate their patents in Italy in spite of the size of the market, because the enforcement procedures are carried out through standard civil courts. Germany, which has the most recognized judges and a special patent court, significantly reduces the risk in these IP assets. Although the strategic IP assets may be more significant, the operational assets are important. Such a criteria is a "fair value" assessment of the legal value.

4.2.2.4 The Recognition of Internally Generated Intangible Assets

For internally generated intangible assets, the business process of arriving at IP assets need to be singled out. However, the strategic element of trust is not covered; only the risk element of future expected benefits is. These processes need to be separated as for acquired intangible assets.

---

[6]For IFRS documents on recognition of internally generated IA see: https://ifrscommunity.com/knowledge-base/recognition-and-cost-of-intangible-assets/#link-acquisition-as-part-of-a-business-combination.

IAS38 provides a framework for (i) recognition of internally generated intangible assets that helps identifying whether and when there is an identifiable asset that will generate expected future economic benefits, and (ii) for determining the cost of the asset reliably. To facilitate this process, IAS38 classifies the generation of the asset into a *research* phase and a *development* phase (IAS 38.51–52).

These accounting principles of dividing assets into "phases" actually provide *some* measure of the dynamic process of R&D. What needs further development, or is missing, is the "IP development, maintenance, use, and trade phase" when it comes to capitalization of these investments as "production assets."

4.2.2.4.1 The Research Phase

> "Research is defined (IAS 38.8) as original and planned investigation undertaken with the prospect of gaining new scientific or technical knowledge and understanding. Examples of research activities are given in paragraph IAS 38.56 and include obtaining new knowledge or searching for alternative solutions. As noted earlier, intangible assets can be generated internally with input from external parties. Expenditures on research or on the research phase of an internal project must be expensed as incurred as an entity cannot demonstrate that an intangible asset exists that will generate probable future economic benefits (IAS 38.54–55)."

4.2.2.4.2 The Development Phase

> "Development is defined (IAS 38.8) as the application of research findings or other knowledge to a plan or design for the production of new or substantially improved materials, devices, products, processes, systems or services before the start of commercial production or use. Expenditures on development or on the development phase of an internal project are recognised as intangible assets if, and only if, an entity can demonstrate all of the following (IAS 38.57):
>
> (a) the technical feasibility of completing the intangible asset so that it will be available for use or sale,
> (b) its intention to complete the intangible asset and use or sell it,
> (c) its ability to use or sell the intangible asset,
> (d) how the intangible asset will generate probable future economic benefits,
> (e) the availability of adequate technical, financial and other resources to complete the development and to use or sell the intangible asset,
> (f) its ability to measure reliably the expenditure attributable to the intangible asset during its development."

From a strategic perspective, additional criteria are needed as to evaluate the level of trust that can be achieved given the strategies adopted (see Chap. 2). The IP assets have a value in reducing uncertainty, not only the direct risk of future economic benefit. These "trust-based" IP assets may need a separate treatment and account or at least specific *standards* of evaluation. Perhaps the proposed four strategies may be a basis for further refinement of categories. These types or categories will help in identifying which strategy firms choose when considering trade in ideas. Knowing the strategies through the IP asset investments allows the performance of the firms to be related not only to the strategies but also to the institutional arrangements – the international trade rules incentivizing to the strategies in the first place by their impact on uncertainty (and risk) – thus evaluating the institutional performance of treaties, a fundamentally important statistic to inform trade policy.

4.2.2.4.3 Goodwill

> "Assets that do not meet separability criterion or contractual-legal criterion are included in *goodwill*. Exclusions include deferred tax assets and goodwill (IAS 38.2–3)."

The IP assets that are not differentiated in the new accounts but still separable will remain as Goodwill.

4.2.2.4.4 Recognition as an Expense

> "When an expenditure on an intangible item does not meet the recognition criteria of IAS 38, it should be expensed as incurred unless it forms part of the goodwill."

Criteria should include strategic IP assets as well as economic ones before being expensed.

4.2.2.4.5 Writing off IP

> "Retirements and disposals of intangible assets mirror the requirements of tangible assets (IAS 38.112–117)."

Here also, criteria should include strategic IP assets as well as economic ones before being expensed. Thus, there would be a writing-off strategic IP and operational IP on separate criteria (and possibly separate accounts).

4.2.2.4.6 In-Process Research and Development

IAS 38.34 specifically requires separate recognition of *acquired* in-process research and development project. As an example, a firm may buy the R&D results from another firm. These are then a separate intangible asset from patents and other IP that may be part of the project.

The standard thus makes the distinction between research and IP, which is of great importance to accounting of trade in ideas based on patent (IP) licensing. Only IP rights granted under some laws are accounted as IP and "pre-IP" intangible assets as such. It allows the focus to be on the value of exchange in IP *during the time there are IP rights to trade*. Activities before IP is granted and after IP is exhausted are not the focus of trade in ideas. However, such extension can easily be made and there is a lot of work in the field such as the Frascati Manual for R&D.[7]

4.2.2.4.7 Traditional Knowledge as Assets

However, not all IP are granted under laws. The animated discussions on whether "traditional knowledge" is to be considered "IP" by the communities "owning" them shows that. It would be important to "make room" for such "community assets" in a note. In a sense, they represent the research results of the communities, and may be personal (held by individuals in the community) but not private as protected by a body of law. See further discussion on the framework. This question is an important issue in developing countries to move from a community system to a private property system where knowledge can be traded through rights.

4.2.2.4.8 Disclosure

> "Disclosure requirements are set out in IAS 38.118–128. IAS 38 requires also explanation of assessment that an asset has indefinite useful life and encourages to disclose *significant intangible assets* controlled by the entity but not recognised as assets because they did not meet the recognition criteria of IAS 38 (IAS 38.128(b)). These may include co-operations with universities, joint ventures, and other assets."

Such disclosures are, in effect, "network" related and not IP related. They represent the *human capital formation* that firms have access to.

A summary of the treatment of intangible assets under IAS is found in Tables 4.2 and 4.3.

A summary of the treatment of intangible assets on the balance sheet is found in Table 4.4.

4.2.3 *Examples of Accounting Practices Today*

As an example of how IFRS15 and IAS38 are applied, we will look at Qualcomm and Ericsson, two of the foremost patent licensing firms in the world. This is based on their respective annual reports of 2021.

The reading makes it very clear how both companies adhere diligently to the principles of these standards and even use the wordings of the standards to describe their accounting practices. It is so close that one might think that the standard was written by these firms. This tells us the state-of-the-art in the accounting practices in relation to revenues from contracts based on patents and other IP.

[7] For the Frascati Manual on R&D Statistics: Guidelines for Collecting and Reporting Data on Research and Experimental Development, see: https://www.oecd.org/sti/inno/frascati-manual.htm.

Table 4.2 The treatment of intangible assets

Financial accounting to measure the performance of management of inventions	
Treatment of intangible assets	Chose in possession focus
Intangible assets	
Intangible assets are non-physical, non-monetary and identifiable (separable or arises from contractual or other rights)	IAS 38.12
Embedded IP	
If an intangible asset is not an *integral part* of the related hardware, it should be accounted for separately	IAS 38.4
Depreciation	
Any intangible asset that will not be "consumed" after one use (like patents, IP), can be treated as an intangible asset with its amortization presented *below* EBITDA, i.e., it is not treated as inventory that is consumed, but a "production asset" (that is "used" and amortized).	IAS 38
Cost	
Under [the] *cost* model, an intangible asset is carried at *cost* less any accumulated amortization and any accumulated impairment losses.	IAS 38.74
Revaluation	
Under the revaluation model, an intangible asset is carried at its fair value (i.e., revalued amount) less any accumulated amortization and any accumulated impairment losses.	IAS 38.72–73
Measurement of acquired assets and liabilities	
The acquirer measures the identifiable assets acquired and the liabilities assumed at their acquisition-date *fair values* (with certain exceptions)	IAS 38.25–32/IFRS 3.18–19
Mergers and acquisitions	
Acquisition as part of a "business combination," i.e., mergers and acquisitions.	IAS 38.33–37
Exchange of intangible assets	
One or more intangible assets may be acquired in exchange for a non-monetary asset or assets, or a combination of monetary and non-monetary assets. The cost of such an intangible asset is measured at fair value unless (a) the exchange transaction lacks commercial substance or (b) the fair value of neither the asset received nor the asset given up is reliably measurable.	IAS 38.45–47
Fair value	
The fair value of an intangible asset is reliably measurable if (a) the variability in the range of reasonable fair value measurements is not significant for that asset or (b) the probabilities of the various estimates within the range can be reasonably assessed and used when measuring fair value. If an entity is able to measure reliably the fair value of either the asset received or the asset given up, then the fair value of the asset given up is used to measure cost unless the fair value of the asset received is more clearly evident.	IAS 38.47

Table 4.3 Recognition of intangible assets

Financial accounting to measure the performance of management of inventions	
Recognition of intangible assets	Chose in possession focus
Recognition of IA	
An intangible asset is recognized when it meets all of the criteria of: *identifiability, probability* of and *control* of the future economic benefits, and *reliable measurement* of cost.	IAS 38.18,21
Recognition of internally generated IA	
The generation of the asset is divided into a *research* phase and a *development* phase.	IAS 38.51–52
Research phase	
Expenditures on research or on the research phase of an internal project must be expensed as incurred as an entity cannot demonstrate that an intangible asset exists that will generate probable future economic benefits. Research projects are thus not considered assets on grounds of uncertainty and expensed.	IAS 38.54–55
Development phase	
An intangible asset arising from development (or from the development phase of an internal project) shall be recognized if, and only if, an entity can demonstrate the asset will generate probable future economic benefits. A number of criteria are set up: technical feasibility, intentions, ability, business plan, sufficient financing to bring to market, and measuring the costs. These criteria are really needed to make the distinction from the research phase.	IAS 38.57–58
Disclosure	
The entity shall make a detailed disclosure of types of IA, IP, life, amortization, and other.	IAS 38.118–128

4.2.3.1 Qualcomm

IP are referred to as "intellectual property." In Note 1 of "significant accounting policies," the revenue sources are described as:

> "We are a global leader in the development and commercialization of foundational technologies for the wireless industry. Our technologies and products are used in mobile devices and other wireless products, including those used in the internet of things (IoT) and automotive systems for connectivity, digital cockpit and advanced driver assistance and automated driving (ADAS/AD). We derive revenues principally from sales of integrated circuit *products* and through the *licensing of our intellectual property, including patents and other rights*." (italics added)

The importance of IP is thus at the center of their revenues. In the income statement, the revenues are "Equipment and services" and "licensing." We can see how the "goods and services" appear, but also "licensing" as a main type of revenue. Two sentences are used to sum up products and services revenues and close to two full pages to discuss the licensing. Licensing is considered a business in its own rights and is introduced as:

Table 4.4 The IP balance sheet

Financial accounting to measure the performance of management of inventions today	
Balance sheet	Chose in possession focus
Non-current assets (long-term assets)	
Capitalized development expenses	IAS 38.8, 57
Goodwill	IAS 38.2–3
Intangible assets	
Cost model	IAS 38.74
An intangible asset is carried at *cost* less any accumulated amortization and any accumulated impairment losses	
Revaluation model	IAS 38.72–73
An intangible asset is carried at its fair value (i.e., revalued amount) less any accumulated amortization and any accumulated impairment losses	
Current assets	
Contract assets	
Remaining use-right to patents that are not yet invoiced. Unclear if applies to only current asset or also non-current (portion >1 year)	
Long-term liabilities	
Current liabilities	
Contract liabilities	
Pre-payments for use of IP, i.e., annual or one-time fixed fee. Unclear if applies only to current liabilities or also non-current (portion >1 year)	
Equity	

> "We grant licenses or otherwise provide *rights to use* portions of our *intellectual property portfolio*, which, among other rights, includes certain patent rights essential to and/or useful in the manufacture, sale or use of certain wireless products. Licensees pay royalties based on their sales of products incorporating or using our licensed intellectual property and may also pay a *fixed license fee* in one or more installments."

The "rights to use" are from a portion of the full "portfolio," and the price is based on royalty on sales and may include a fixed fee, i.e., a "linear contract."

On the asset side, "Completed technology-based intangible assets" are about 12% of "Goodwill." "In-process research and development" is about 11% of "Goodwill." These categories of intangible assets represent 61% of total assets. The language used thus strictly adheres to IFRS15 and IAS38. Reference Qualcomm Annual Report 2022 (2022).

4.2.3.2 Ericsson

IP are referred to as "Intellectual Property Rights" (IPR) by Ericsson and is seen as a separate "patent and licensing business": "This type of contract relates to the patent and licensing business."

The nature of the IPR contracts is such that they provide customers a license with the right to access the Company intellectual properties *over time*, therefore revenue shall be recognized over the duration of the contract.

Royalty revenue based on sales or usage is recognized when the sales and usage occur. The transaction price on these contracts is usually structured as a royalty fee based on sales or usage over the period, measured on a quarterly basis. This results in a receivable balance if the billing is performed the following quarter after measurement. Some contracts include lump-sum amounts, payable either up-front at commencement or on an annual basis.

Revenue from IPR licensing contracts is allocated to the *segments* Networks and Digital Services (Note B1: "Segment Information"). Reference Ericsson Annual Report 2021 (2021).

4.2.4 The Need for a New Accounting Definition of "Product"

Replacing a "chose in possession" with a "chose in action" (for all but a limited number of products?) may be a way to create an accounting system that is more suited to the idea-based economy of the twenty-first century. The current standard builds on physical goods that are physically transferred. This standard is then expanded to services but now increasingly licensing within IP-intense industries. From an economic and business point of view, these are mere rights, not things. The "fetish" approach to trade needs to be replaced by a "rights" approach to trade. What is "transferred" – to recognize revenues – are the rights to use, reuse, redesign, recycle, and re-trade this bundle of rights. When a physical good is transferred, there are some limitations to what you are allowed to do, for example repairing (warranty), reusing (environmental concerns), and refinancing (age). The same goes for services.

4.2.5 The Need for New Accounting Tools

New software tools are also needed for tracking and maintaining contract revenues, asset, liabilities, depreciating, and revaluing and capitalizing IP assets.

4.2.6 Patent (IP) Assets and Liabilities on the Balance Sheet

Today's licensing market is much more of a sellers' market – where patent holders seek out possible users and infringers of the rights – than a buyers' market – where buyers first check out what patents are out there and seek a license. Both exist, of course, but for example in the telecom business, it is common that the patent holders of the standard essential patents (SEP) do not initially seek a license from an infringer until they have become a more important player. This means that a (large) portion of the goods and services market development is done under de facto zero-royalty licensing, albeit an implicit one. In order to understand the value flow based on patents (IP), it is important to capture these implicit zero-royalty agreements as well. That requires a differentiation on the de facto use of IP owned by others (liability side) and the ownership of IP (asset side) as these implicit agreements are not disclosed.

New *accounting standards* are needed for activation, depreciation, and periodic revaluation of IP assets if the value of trade flows from ideas (IP) and IP asset positions (assets and liabilities) should be measured. A new international patent position statement (IPP) is proposed to publicize the ownership of the patented technology in use. It will show the origin of the patents in use in a given country and which countries use the patents originating from that country's inventors. It is thus similar to the international investment position (IIP) where the international lending and borrowing are summaries. Here the ownership of the in-licensing and out-licensing of patents are recorded to create a "patent position statement". See Sect. 5.3.

An important aspect when licensing is that you can only license assets that you own and know that they are (presumed or deemed) valid. Thus, a license contract by a licensee represents a *liability* on the balance sheet and an *asset* for the licensor. The best way to think about this is perhaps that patents (IP) are *production assets* that can be owned, in-/out-licensed, or even "leased." This is a relatively new, if not completely new, issue in accounting. The liabilities are balanced by the IP assets owned and the contracted rights, an asset that is depreciated similarly to a fixed asset. Treating IP assets as "production rights" means that the "technology factor" (as in international economics) becomes measurable at the level of the firm.

4.2.7 Summary of Recognition of IP Assets

In summary, we see, as for recognition of revenues, that the principal starting point in the recognition of intangible assets is tangible asset. A sequential process is the "model": (1) research and (2) development. If we treat property rights more generally as a bundle of rights, which they more generally are (there are also limitations to the use of tangible rights), then this bundle may or may not include IP. Thus, the simplest product can be IP-only, and more complex ones, goods and services.

4.2.8 Summary of Recognition of Revenues from IP and of IP Assets

The whole idea seems to be that the revenues are recognized when the IP is accessed or used, i.e., *producing* something for the customer. This "production" can also mean no production, i.e., accessed but not currently used in the production of a good or service.

Since "sitting on" the right is a common strategy to "block" others from using the IP at all or timing own use with goods or services upgrades, such as an option, seller/licensees try to negotiate a one-time or periodic fixed fee together with a royalty based on actual use. This is to give incentives for the agent to use the IP, resolving the principal–agent problem in favor of the principal. For an experimental economic study on why this may be the case, see Ullberg (2012, Ch 3.5, p. 114).

The statistically significant result of this study shows that, in a competitive market with multiple buyers of patented technology, buyers are willing to pay the value of the second highest "blocking use" to acquire the rights. The value of blocking is thus revealed. The expected result would be that buyers bid to zero but instead they appear to "ensure" the value of current goods and services by blocking a competitor's use of the IP.

According to IFRS15, the fixed fee would be recognized when the license is granted and the royalty part when the IP is used by the customer.

From a trade in ideas perspective, the current treatment is thus taking the starting point in the production and sale of goods and services that may use IP rights in their production. The rights may be embedded in the goods and services and sold as a "bundle" or may be sold or non-embedded and separate license.

To deal with the aspects of embeddedness and non-embeddedness for licensing of intellectual property, a distinction has been made between whether the sale of a *good* or *service* that is "distinct" or not. When there is a bundle, but the sale is *predominantly* related to IP, that is also considered distinct.

Either the recognition of revenues is as an IP good or service or as a "bundle," but there is no separation of the IP from the bundle. Sometimes it is possible that there are two "performance obligations" involved in contracting: one for the good or service without the IP value and the other for the IP value. Since it is difficult to find a way to separate the two in a single "performance obligation," it is not recommended that a separation is made ad hoc.

A sale of IP is more straightforward and recorded as a separate good or service if it is [predominantly] "distinct." The definition of "predominantly" is somewhat challenging.

For statistics purposes, the concept of trade in ideas (IP) as a *product* in financial accounting is consistent with the proposed balance-a-payment principle of trade in goods, services, and ideas in three current accounts.

In financial accounting, establishing *similar* principles for the transfer of promised ideas (IP) to the customer in an amount that reflects the consideration to which the entity expects to be entitled – through transfers or licensing/cross-licensing – in exchange for those ideas (IP) will provide the data source for such statistics. As cross-licensing is perhaps the most common licensing, the "exchange" of thus both in monetary terms ("the value") and in other IP results in a change in the balance sheet on the IP asset side (licensor) and liability side (licensee), as well as financial assets for payments. This recognition of IP as assets on the *balance sheet* – debated for decades – appears to be a completely necessary part of a transaction related to trade in ideas. See next section.

The recognition of revenues is typically made over time of the license or in a combination of a fixed one-time or annual fee and a royalty (which can have a minimal level, such as a fixed fee, vary in size over some volume, or be based on sales or profits). This risk-*sharing* mechanism is common as one cannot predict future sales with certainty. It also gives the licensor the option of non-use, "blocking" others from using the rights, which is a very common strategy. The recognition for the certain part – fixed fees – is done at the signing of the contract and the recognition of revenues from the royalty part after their use. However, because the actual use cannot be known until a report is completed, the revenues may have to be adjusted after an estimate is made. All these current rules make sense in terms of realizing the *economic value* of their use, closely following the value added to the economic system.

It is now high time to define the data gap and structural differences between current standards and the proposed theory.

4.3 The Data Gap Compared to the Proposed Theory of Value

To address the data gap, we need to further differentiate the revenue streams and asset and liabilities compared to current standards. We also need to have more data on the risk and uncertainty in the contracts for these highly uncertain assets, which are largely incomplete.

4.3.1 The Provenance of Revenues – New Definition of Revenue Types

To separate out the revenue streams from patent transfers and licensing and other IP from those of goods and services, the revenues need to be grouped into "types" of revenues based on their *provenance*, i.e., relating them to the asset from which they were generated. The concept of provenance is not in the current standard. As a matter of fact, it appears that there is no definition of "goods" and "services" either.[8] The assets delivering the revenues should be grouped into IP assets and liabilities for "ideas," tangible assets for "goods," and human capital for "services." We thus need information on these three revenue streams and these groups of assets.

The idea is that the best way to view property rights, which a firm can own, is as a *bundle of rights*, as in fact the rights are not free for all but rather constrained when it comes to their *use*. This division furthermore captures the dynamic dimension of

[8] The definition of "goods and services" is not in international financial accounting standards. Personal communication with Baruch Lev, accounting prof. at NYU Stern.

development, production, and use/reuse, by introducing the time dimension discussed earlier. The technology and ideas are first developed and traded, then marketable goods are developed 10–20 years later, and finally these goods are serviced at the customer's end to perform what they are used for. One can thus measure the numerator (returns) in the performance of management measure of trade in ideas that we are interested in, RIPA.

4.3.2 A New Structure to Measure the Return on IP Assets

We are interested in measuring the performance of the management of trade in ideas for a company, and the IP assets that a firm has *access* to, i.e., controls or has residual right to control of (as in previous chapters always focusing on patents but expandable to present and future IP rights).

Other measures of the management of the invention process are of interest, but the Return on IP Assets, RIPA, is an ultimate measure of a firm's performance in this line of business: how much profits can be generated using the firm's IP assets.

To this end, we need to define these assets. The view here is that property rights in general ought to be viewed as bundles of rights. This follows Coase (1960)[9] viewing factors of production as rights. As patents are granted for a certain *technology*, have legally defined *field of use* (claims), are validated in a particular *geographic* jurisdiction, and are valid for a certain *time*, the rights can be seen as a bundle. In addition, they are transferrable and licensable with mutual consent, are often themselves bundled in patent pools, or cross--licensed for other bundles of rights for other markets, makes a "bundle" both economically and businesswise meaningful in terms of what you can use these rights for. Similarly, goods and services have limiting rights when it comes to their use. Consequentially, *all production assets* can be viewed as bundles of rights, separating them by "type" – intellectual (and other) use rights, tangible assets (use rights), and human capital (right to the fruit of people's knowledge).

By categorizing the revenues by provenance, we can measure the performance of management in ideas, goods, and services. The purpose here is not to discuss goods and services, but a principle to separate out the returns from IP from other production assets in order to calculate RIPA. This principle also established the view of IP rights as *production assets* and liabilities.

When it comes to the contract, all contracts are incomplete in the sense that not all contingencies can be specified.[10] This is much a business cultural issue, whether the business communities "honor the inventors" or not. Thus, data on the business culture of the contracting parties needs to be collected to account for the value of

[9]See Coase (1960, p. 44).

[10]See Hart (1988).

these rights. It is not merely a manner of legal regime, but also how it is implemented in practice.

Thus, the following additional data are needed:

- The revenues divided into three types of products based on their provenance: *ideas*, goods and services.
- The IP rights and contracts on IP rights as *assets.*
- The contracts on IP rights as *liabilities* (for the licensor).
- The *business culture* of the contracting parties (on a scale honoring the inventor).
- The *ownership* of the underlying rights (patent, IP).

Since we want to do this in a standardized way – to create comparative data and statistics – we need to define the data gap to the current standard. Complementing the current standard with this data, we get the following gap analysis table. See Table 4.5 for income statement and Table 4.6 for the balance sheet. The additional information on culture and ownership needs to be sourced through surveys or other databases on this data.

This section has outlined the *data gap* and attempted to place the required data in an expanded and restructured version of the current standard. The next chapter will outline the proposed framework based on the principles for data collection and the practical considerations for a discussion on new standards.

Table 4.5 Accounting data gap

Financial accounting to measure the performance of management of inventions today		
Income statement	Chose in possession focus	Chose in action focus
Revenues		
Principle: Recognition of revenues comes from contracts on goods and services which have single or multiple performance obligations with customers (often disaggregated by entity's operating/ market segment)	IFRS15.1–2	Expanded
Recognition of revenues comes from contracts on ideas, good, and services (disaggregated by the entity's bundle of rights to productive assets, technology, operating/market segment)		x
Promised goods and services include the granting of licenses of IP	IFRS15.26(i)	Expanded
Promised ideas (IP), goods, and services are separated by provenance (the type of assets used for their production)		x
Ideas, goods, and services		Expanded
Contracts where IP is not a primary performance obligation	IFRS15	Refined
Contracts where IP is not included in any performance obligation (open access)		x
Contracts where IP is the only performance obligation		x
Licensing of IP		

(continued)

Table 4.5 (continued)

Financial accounting to measure the performance of management of inventions today		
Income statement	Chose in possession focus	Chose in action focus
Risk adjusted licensing revenues from contracts with distinct, or treated as a single, performance obligation, sometimes disaggregated by use segment, based on portions of an entity's IP assets		x
Contracts may include IP updates over the time of the contract (a "capture period") or for specific IP or cross-licensing with net payments		x
Distinct (non-embedded, non-embodied, un-bundled) IP *only*	IFRS15.B52–53	
At a point in time (such as a fixed fee or royalty)		
Over time (such as annual fixed fee or royalty based on use)		
Non-distinct (embedded, embodied, bundled) where IP is the *primary* component of the performance obligation	IFRS15.B54	
At a point in time (such as a fixed fee or royalty)		
Over time (such as annual fixed fee or royalty based on use)		
Transfer of IP		
Sale and purchase of IP; affects IP assets on balance sheet	IAS38	
Exchange of IP	IAS38	
Exchange of IP asses without any payments; affects IP assets on balance sheet; a net-payment may be part of transaction		x
Cost		
Research phase		
Research phase: Research costs are sometimes depreciated but mostly expensed	IFRS15	
Development phase (IP)		
Recognition of cost under a cost model or a revaluation model	IAS 38.72–73	
Maintenance phase (IP)		Refinement
Recognition of costs of IP renewal fees, challenge, enforcement		x
EBITDA		
Depreciation: amortization and any impairment losses	IAS 38.74	
Intangible assets are depreciated for their economic life, i.e., 10 years		
Research costs are depreciated over shorter periods, i.e., 3 years		
EBIT		
Financing with IP collaterals		
EBT		
Tax on IP		
EAT		
Income from IP		

Table 4.6 Accounting standard gap

Financial accounting to measure the performance of management of inventions today		
Balance sheet	Chose in possession focus	Chose in action focus
Non-current assets (long-term assets)		
Capitalized development expenses	IAS 38.8, 57	
Goodwill	IAS 38.2–3	
Intangible assets		
Cost model	IAS 38.74	**Refinement to strategic assets**
An intangible asset is carried at *cost* less any accumulated amortization and any accumulated impairment losses		
Revaluation model	IAS 38.72–73	
An intangible asset is carried at its fair value (i.e., revalued amount) less any accumulated amortization and any accumulated impairment losses		**Refinement to strategic assets**
Current assets		
Contract assets		
Remaining use-right to patents that are not yet invoiced. Unclear if applies to only current asset or also non-current (portion >1 year)		**Contracts on IP as assets**
Long-term liabilities		
Current liabilities		
Contract liabilities		
Pre-payments for use of IP, i.e., annual or one-time fixed fee. Unclear if applies only to current liabilities or also non-current (portion >1 year)		**Contracts on IP as liabilities**
Equity		

Part III
The Statistical Framework for Trade in Ideas

The overview of the statistical framework for standards and trade in ideas is first outlined at the level of principle and measures, followed by a discussion/proposal of changes in principles of the current system, and how each element could be implemented through current and new standards. (The new standards will follow from the principles and will be developed through international standards organizations. An outline of this process is proposed in the last chapter.) Finally, a proposal for expanding financial accounting and balance of payment *measures*, including a new trade in ideas *index*, is presented in summary form.

Chapter 5
Overview of the Framework

The statistical framework for trade in ideas builds on the fact that trade *begins* with the ideas, which are then used in goods and then services, or trade is *first of all in ideas.* This sequence forms the logical description of the economic system used in the proposed framework. The most important ideas are the technical ideas as they provide productivity gains for economic growth. "Ideas" refer to the patents and other current and future intellectual property rights granted for the inventions. By separating business activities in this way, i.e., by the *time* when they occur, allows the *dynamism* of the economic system – the interaction of ideas, goods, and services over time for a more productive outcome – to be better captured in the statistics. This is quite similar to the current logical separation of goods and services as services typically emanate from increasing the performance of the development, use, and reuse of goods.

The statistics for the highly dynamic economic system of the twenty-first century will then bring the activities with the highest economic returns to the forefront, the ideas. The ideas and the trade in ideas are a key measure of the leverage of the human capital formation of nations, which is at an all-time-high. This is especially true for the developing nations. See Fig. 5.1.

The statistical framework presented captures this dynamism by expanding the data for economic analysis to:

(a) an *economic* environment where (i) the *revenues* and (ii) *assets* from the idea production are separated out from other revenues and assets, and (iii) the *strategies* used by idea-producing entities to trade IP and use IP are identified.
(b) an *institutional* environment of (i) *rules and norms*, which may be formal or informal and structures the trade in ideas; (ii) the *contracts*, which is incomplete and may cover certain technology areas, field of use, markets, time, and other constraints; (iii) the *prices*, i.e., the value, specified by the contract dimensions created by regional and international trade and specialization in ideas; and (iv) the *payments*, which are made in money and/or other IP and other assets. Payments are not related to revenues (the economic environment) but to the

E. Ullberg, *Intellectual Property Statistics*, Contributions to Economics,
https://doi.org/10.1007/978-3-031-36386-3_5

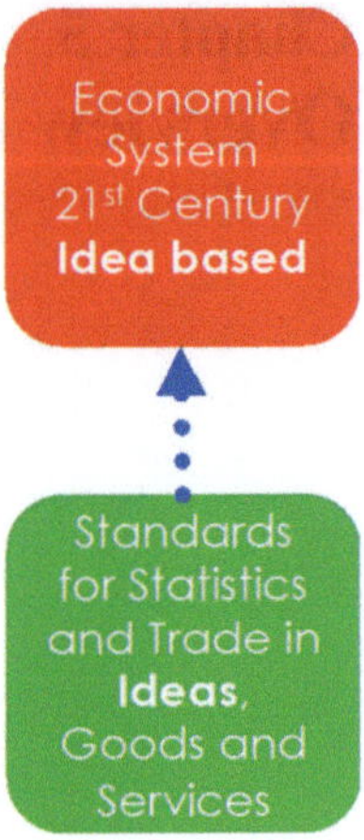

Fig. 5.1 The statistical framework aims at capturing the dynamism of an idea-based economic system through expanded and differentiated standards for data collection, how data are compiled, and new measures for trade in ideas calculated

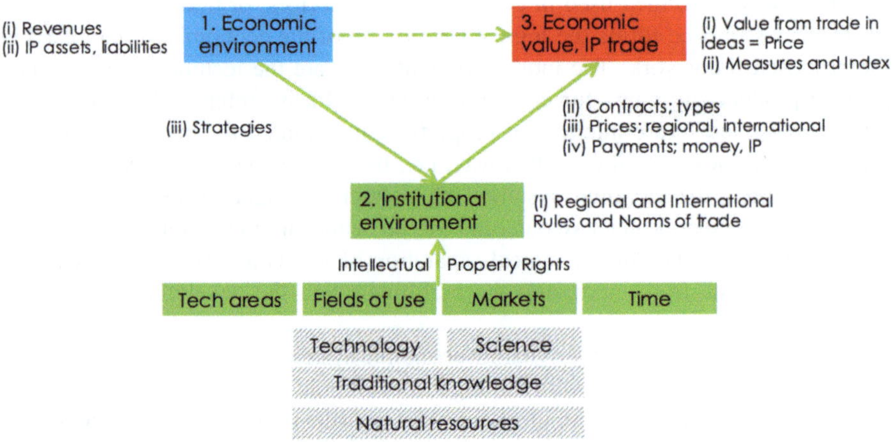

Fig. 5.2 The idea-based economic system used for the statistics framework

exchange of assets to consummate a transaction (the institutional environment), as in current accounting standards and economic systems theory.

(c) The *dynamic* economic outcome is thus a *value* from trade flows in ideas, which is dependent on the economic environment creating the revenues *given* the strategies used by agents, in turn, *given* the institutional rules. The statistical separation of the strategies allows the measurement of the performance of the *management of inventions*. The separation of institutions allows the measurement of the institutional performance and behavioral properties of *markets* in patents/IP, governed by regional and international rules (treaties, agreements, standards, etc.) and norms (see Fig. 5.2).

We will now briefly look at the elements of the dynamic economic system for trade in ideas and the proposed statistical principles for the expanded data to be collected. Then we will discuss the measures made possible through the new data and relate the current standards to the expanded framework. Finally, we will briefly discuss the formal process to adjust the statistics for measuring and standards for trade in ideas.

5.1 The Economic Environment

5.1.1 IP Revenues and their IP Asset Provenance

The central *principle* for characterizing IP revenues is that the *revenues* from cross-border trade can be categorized by the *type of assets* used in producing the things sold, their *provenance*,[1] and that these assets are, in turn, directly related to the *logic* by which the ideas, goods, and services are developed and traded. The assets used to produce idea (IP) revenues are IP asset, physical assets are used to produce goods, and knowledge (human capital), information, and systems assets that the firm own or have access to are used to perform services. This means that patent and other IP as well as contracts on these rights are to be considered tradable *production assets* and *liabilities*.

A licensor offers contracts based on IP assets or other contracted rights. This creates a liability (a promise) to deliver these rights over the contracted time. A licensee accepts a contracted right, which creates a production rights asset over the time of the contract.

The revenues can then be categorized as *ideas, good, and services* products (commodities, contracts), each produced using a "bundle of rights." Compare Coase on how property rights are best characterized as "bundles of rights" as all rights have constraints on what they can be *used* for. These "bundles of rights" are in essence personal (such as knowledge) and impersonal property rights on things in possession and things in action (see Fig. 5.3).

The logic – business or economic – is decided by the way these assets as used and activities are *managed*: creation and use of rights (IP) as assets; purchase/construction and use of physical (tangible) assets; and access to and use of knowledge (human capital formation), information (data), and systems as assets. However, IP intense firms have added "revenues from IP licensing" as an information line under "revenues from goods and services," *recognizing* the importance of IP as a source of

[1] This is a new concept to attempt to explain the economic rationale behind dividing revenues into types: ideas, goods and services. In the current standards of IFRS15 and IAS38 there appear to be no formal definition of "goods and services." The term "goods and services" appear instead to be taken from business practice, what the firms call that they are selling. The word "provenance" is taken from the world of arts, used to describe the ownership history. Here the word is used to refer to the assets that are used in creating the product.

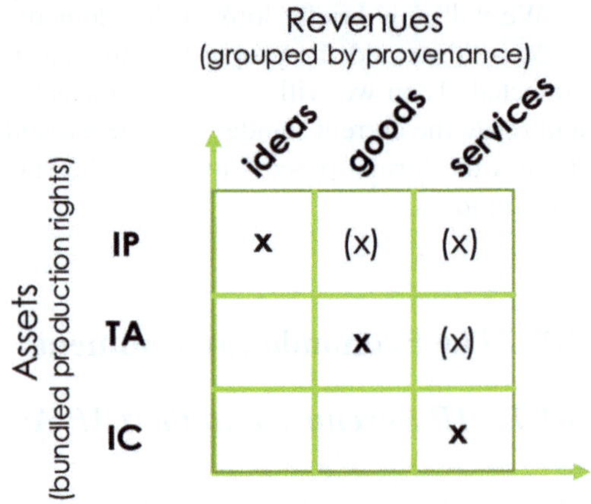

Fig. 5.3 Revenues are classified according to their provenance (assets used in their production)

their revenues and income (profits). In this way, Return on Assets (ROA) can be calculated and returns from different assets compared and evaluating management on Return on IP assets (RIPA).

The "stock of technology" can also be estimated through the IP assets and IP contracts.

Among other things, a new, separate, *current account* is proposed for trade in ideas in the balance of payments, differentiating revenues from trade in goods and trade in services.

5.1.2 IP Assets and Liabilities

A second *principle* for measuring the economic returns on ideas is that patents and other Intellectual Property Rights are considered as *production assets* in a company's *financial* accounting, in *national* accounting, and in *international accounting*.

Development costs (from ideas to IP rights) and cost of purchase and sale are accumulated (capitalized) and then depreciated according to new standards (outlined here but need to be formally developed by standards bodies). This allows the calculation of RIPA, return on patent/IP assets at the firm and national levels, and the leverage of human capital formation of firms and nations through trade in ideas, measured by terms of trade (for each technology area) and an international patent position (IPP), stating ownership (national/foreign) of the IP in use (accessed).

Both IP as assets and liabilities and ideas as revenues are an expansion from current accounting and balance of payment standards, which classify revenues as "goods and services" and have only minimal capitalization of IP assets.

For example, today's accounting standards of Goodwill and capitalization of some research costs and software in some jurisdictions are expanded to *all* IP rights – present and future – that have future revenue-generating value.

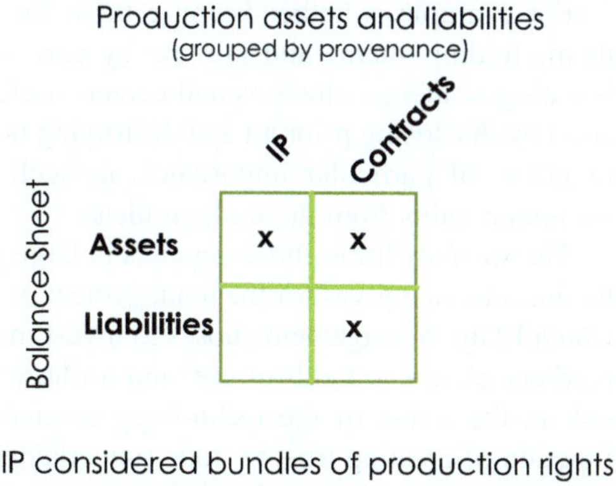

Fig. 5.4 Access to IP as production rights are considered assets (for licensee and licensor) and liabilities (for licensor)

Simple rules should be developed where the principle is to accumulate all costs in the development of IP. Annual revaluation of these assets with respect to their potential revenue-generating value should be done. New accounts should be created for these assets and their depreciation and write-off to keep them separated.

5.1.2.1 Contracts as Assets and Liabilities

A third principle is that *contracts* on IP are also considered *assets* for the licensee and *liabilities* for the licensor based on the principle that they are production assets for the firms. It is access to the production assets that counts for production, now ownership. The value is capitalized during the contracts' economic life both as assets (the licensee) and liabilities (the licensor) (see Fig. 5.4).

5.1.2.2 Incomplete Contracts and Residual Rights

Since contracts are by definition *incomplete* (one cannot take into account all contingencies), data on the resolution of such conflicts are important to include in the valuation, i.e., the enforcement of contracts. Social norms, culture, and legal rights with respect to honoring the inventor will be measured. These measures are seen as part of the institutional environment, and thus affect the choice of strategy and the economic outcomes.

5.1.3 Strategies

To understand the value created through the ideas – patents and other IP – we need to know how and why firms actually use them, i.e., their *strategies* and, in turn, whether these strategies are also *economically* efficient.

For example, whether keeping them for own use, blocking others from using them, trading them, abusing the system with frivolous lawsuits, or other like boosting academic citations and careers, defensive publishing for freedom of operation by disclosing prior art and destroying novelty. Why firms choose to trade them or not is of particular importance, as such exchange leads to specialization and economic gains from the trade in ideas.

The *strategy* firms chose depends in large part on the institutional framework and the incentives it gives for the management of risk and uncertainty. Strategies refer to *which* IP the management choses to invest in (which technology areas, jurisdictions/ markets, claims or fields of use, and for how long) to reduce the uncertainty and the risk in the value of the technology to enable exchange. The IP assets have the function of creating trust to trade (uncertainty) and creating assets to trade (risk).

What is traded is thus decided by these strategies. The value (the revenues) of the trade flows from ideas can then be connected to the strategies chosen and the institutional framework in place. Mathematically, this dual–conditional relationship can be expressed as Equation 5.1 (compare Smith, 1982).

Equation 5.1 The function of economic value from trade flows from ideas

$$v = f(e,\ S|I)$$

e = agent (inventor, company, university), agent characteristics (technology, knowledge, research budgets, and its preferences and culture), and idea products, S = strategy, I = institutional rules of trade, and contracts transacted.

Since the idea-based economy is a highly dynamic economic system as described earlier, the incentives for one strategy over another may be induced by the institutional trade rules. We thus need to measure the process by which firms arrive at and change their strategies given the incentives of actual *trade agreements* (free trade and other).

5.1.3.1 Messages of Negotiating a Strategy and Contracts

In Chap. 1, we discussed how and why firms choose a certain strategy for trade in ideas. In describing the idea-based *dynamic* economic system, we are concerned with the "messages" firms send to one another *preceding* the final allocation-determining "messages" as in a negotiation. These messages are of two types: (1) "messages" to arrive at a strategy that *reduces uncertainty* in the value of patents and IP intended to be traded (or not traded) in the future of which some are not yet invented to a level where rational estimates of a (future) value can be made, and (2) "messages" to negotiate the contract on IP rights to use in idea-bundles, goods, and services.

5.1.3.1.1 Strategic Response Messages

Since *strategic response* is decided in response to *all other agents' strategies*, the "negotiation" of which strategies firms chose is an iterative process over time. The first messages are investing in (publicly visible) IP positions (assets), *signaling* which strategy is used to either "stay clear" or "reciprocate." See Equation 5.2 for a mathematical description of the first process. These strategies therefore need to be made "visible" through (official) financial accounting and statistics to reduce costs for all inventors to identify them. Today's markets are plagued by hide-and-seek behavior using the patent system classification codes, shell companies to hide ownership or "junk" patents increasing search costs for quality patents (with potential economic value). This is major institutional policy area related to transparency. These "search costs" likely need to come down several orders of magnitude for a level playing field for inventors of all sizes.

Equation 5.2 Strategic response behavior of agents

$$S^i(t) = f^i\left(S(t-1)|e^i,\ I\right)$$

$S^i(t)$ = strategy of agent i, $S(t-1)$ = strategy of all agents at t–1, e^i = agent, agent characteristics and idea products of agent i (developed, purchased or licensed), I = institutional rules of trade (for IP or contracts on IP). Changes in e and I will thus impact the choice of S by agents in addition to changes in S by other agents.

5.1.3.1.2 Contract Negotiation Messages

The second type of messages are (private or public) offers, as in a contract negotiation, which leads to a (private or public) price. Here the institutional rules may be formal or informal and ends with a formal (incomplete) contract agreement, whose revenues are accounted for in the financial accounts, or no contract agreement. See Equation 5.3 (compare Smith, 1982).

Equation 5.3 Messages exchanged preceding final allocating message

$$m^i(t) = f^i(m(t-1)|e^i,\ S^i\ |I)$$

$m^i(t)$ = offer of agent i at sequence point t, $m(t-1)$ = earlier offers of all agents, e^i = agent, agent characteristics, and idea products of agent i (developed, purchased, or licensed), S^i = strategy of agent i, I = institutional rules of trade (for IP or contracts on IP). Changes in e or S by other agents or I by new treaties will thus impact m.

This response behavior follows the investment strategies of all firms and the agent's particular characteristics sustaining their business model, e.g., idea, good, or services focus.

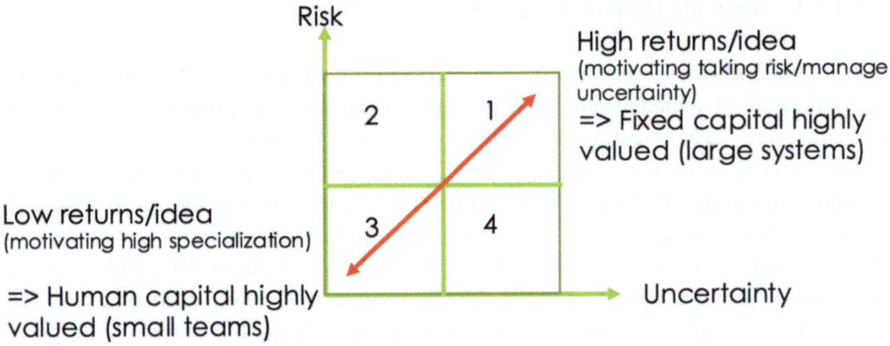

Fig. 5.5 Four strategies characterize firms' investment to manage uncertainty and trade behavior to manage risk in ideas: (1) staying clear, (2) strategic alignment, (3) marginal contracting, (4) systemic abuse

The classification of strategies is based on how firms manage both risk and uncertainty in the value (revenues) from IP, and is from a study of the most patent licensing active firms in the world referenced earlier (Coase, 1960; Ullberg, 2015a). As discussed in Chap. 2, they are as follows: (1) staying clear, (2) strategic alignment, (3) marginal contracting, and (4) systemic abuse (see Fig. 5.5).

Strategies 2 and 3 result in higher specialization and competition that the last two are economically and socially preferable. This is because the selection of ideas to invest in is more efficient in markets with *prices* – whether public or private – as managers know what the value is (what people are willing to pay for the ideas).

The strategies that result in more trade in ideas result in higher economic performance as specialization takes place between inventive firms. This is also because not only business opportunities with super-high returns can be chosen but also the many lower return opportunities, distributing the benefits of the invented technology through international markets and increasing their economic use. As a result, these larger markets further encourage specialization as the market risk is even lower. The super-high return requires massive investments (capital) to develop and realize them. This is only a game for the powerful few large nations. The lower return opportunities are accessible for a much broader set of inventors, and inventor firms, especially important – and readily accessible – for developing nations.

5.1.3.2 Identification of Strategies and Management Performance

To identify firms' strategies and their management's performance, data is collected on the way they manage risk and uncertainty. This is seen in the type of IP assets invested in and the type of contracts used. For example, investments made in certain technology areas, markets, field of use, enforcement, and so on, a balance sheet item, and the type of contracts made (transfers, cross-licensing, licensing, etc.), an income as well as balance sheet item.

The classification of the strategies pursued by IP-intense agents and their investments are therefore the data needed to understand the strategies firms use or are aiming to achieve, given the institutional rules.

One can thus measure the strategic response behavior of all firms by collecting data on the investments.

One can measure the institutional performance and behavioral properties (in which strategies are used) by collecting data on regional and international trade rules.

The framework extends the accounting data to *IP assets and liabilities*, differentiation of *idea revenues*, IP *classification data*, IP *ownership data*, *surveys* (as public IP ownership data is very incomplete and not up to date and firms tend to share data in surveys as long as it remains anonymous), investment strategies, and international free trade agreements.

The four generic strategies are used to categorize the *business models* used (IP trade, IP licensing (incl. cross-licensing), trading the goods and services embodying the IP, or rent seeking abuse of the IP system).

5.1.3.3 A Note on Licensing Strategies in Practice

In many cases, firms are infringing such rights willfully, unknowingly, and yet other times it is unclear. Owners of IP may also "accept" such use as a strategy: let the companies grow and then, when they are big, you ask for a license. These are in effect implicitly granted zero-royalty licenses in the first stage where the demand side sets the price (= 0). Only later the supply side enter the negotiation to create a mutually agreed price.

This behavior is contrary to standard economic theory, where patent holders are considered "temporary monopolists," and skim off all the profits from the users by holding back supply while the users reveal 100% of their demand. Instead, the strategy appears to be to induce adoption of a technology – which has network effects – and then try to get a deal with the larger users to recover the cost. By not honoring the inventors, the infringers risk being run out of business or forced to develop own, protected, technology or stick to trade secrets. The strategy is covered by strategy 4 (systemic abuse), which may turn into strategy 3 (marginal contracting), 2 (strategic alignment), or 4 (staying clear).

Additional information on ownership and usage is therefore needed to calculate a more accurate return on firm and national levels. A complete and updated and accurate registry over rights is needed (this is not the case today where the official IP registries seem hopelessly behind) and holders need to report such infringements. This is necessary to calculate the returns on the IP created by a firm or country, even if the "distribution" of the technology is illegal and benefits other firms and countries without honoring the inventor firm or country. The returns may thus be much higher than the firm can "capture" through the rights. However, with such distribution, there is no exchange and specialization taking place between the producer of technology and the user.

The returns should be calculated on the total use of the technology, distributed through markets with contracts. Including informal contracts is therefore a way to measure the returns on protected technology when network effects are an important driver in its economic use.

5.2 Institutions and Contracts

5.2.1 Institutional Rules of Trade

Institutional policy greatly impacts the strategy chosen by firms. The agreed norms and rules of trade, the quality of the enforcement mechanism of contracts, and taxes may reduce the risk and uncertainty firms face when trading ideas. The economically desired behavior of firms would be more exchange and specialization to raise the gain for all. The data needed is therefore a characterization of these institutions so that the outcome can be related to specific rules of trade. The proposal is to use the free trade agreements (FTA) with regional and international enforcement mechanisms (e.g., WTO) and tease out how they relate to IP as tradable rights.[2]

The institutional performance and behavioral properties of trade rules for cross-border trade in ideas are measured through the level of *specialization* that takes place in the development of technology. A major work is ahead to discover the institutional performance and behavioral properties of markets in patents given these rules.[3]

The framework is designed to allow economic performance data to be related to the institutions in use. The business and economic returns depend on these rules, as well as the time it takes for prices to converge, i.e., price discovery of which technology may in fact be the most valuable.

A fundamental *principle* to characterize the FTAs is whether the rules "honor the inventor" or not. Not only the businesses, not only the society, but also the inventor. The principle implies that inventor firms of all sizes must have access to the use of IP rights (to disclose, protect, defend, trade or own use). This is a complex institutional design goal in its own rights for the patent (IP) systems (offices and courts). To be accessible to all sizes of firms is important as inventions are typically made by individuals or small teams, some who may be linked to an SME, a some to a mid-sized university or a large MNC. See Ch. 1. **The access to the trade system based on IP rights therefore needs to be a level playing field for all inventors, to turn the economic desert into a blooming valley**. This balance between the inventors' rights, the firms use, and the society at large must accept highly asymmetric gains for the inventor (who bears the brunt of the risk and uncertainty), followed by the firms (risk) and then society (consumption). To this end the taxation policy is important so that low-risk activities (such as consumption) are taxed, and the inventive high-risk activities are less taxed or exempt.

---

[2]This is much more than the TRIPS agreement, which focuses on the "minimum standards" of the granting procedure of the IP rights used in trading goods and services, not explicitly the rules of trading the ideas themselves.

[3]At this stage of international trade, the topic of "trade rules for IP" appears largely "unknown" or understudied at policy levels including international organizations such as WTO. Mainly the legal aspects (the rules leading to the establishment of IP rights) have been given a thorough analysis. This has been the case, it seems, at least since the nineteenth century. The view of layers was that once the economic policy was sorted out, the job of implementing it – the trade rules – should be the job of lawyers. This has weakened the economic understanding of the patent system, notably as a *trade system*, which thus allows for specialization and gains from trade. Consequently, no such institutional provisions appear to be in the statistics.

Data on classified institutions (for trade) and taxation policy in relation to the inventive process (inventive firms) are used to characterize these institutional policies. Such data may not be part of balance of payments or financial accounts but need to be collected in addition to evaluate the institutional performance.

5.2.2 *Contracts*

Contracts are typically non-standardized when it comes to licensing of IP and transfer of IP. Monthlong or yearlong negotiations typically preceded such contracts and prices. However, as a matter of practicality, the contract become quite similar in the end. A principle here is to distinguish between a few types of contracts: licensing, cross-licensing in exchange for money and/or in exchange for other IP or technical knowledge/know-how.

The "basic" contract is a two-part tariff contract – a "linear contract" – with one fixed component and one royalty component. These are of different types: fixed fee + royally, fixed only (lumpsum), royalty only, zero-royalty. The assets used for payments are money (currency) and other IP, which should be classified by type of IP used in exchange for the contracted IP. Cross-licensing contracts are then net-payments. There are some standard contracts such as FRAND types of contracts in the telecom industry, which have specific patents listed, the ownership of the rights, and nondiscriminatory license clause.

The IP accessed should also be accounted for by technology area, type of IP, to be capitalized. This includes contracted IP. Such data is typically private, but through surveys types of technology transacted or accessed or out-licensed should be separated on the balance sheet by recognized codes such as the International Patent Classification codes.

The principle is that the contracts should reflect the technology areas contracted and transacted. This allows to measure any specialization of technology areas over time. It also allows to measure where a nations technology is being used, i.e., how broadly it is distributed.

The incompleteness of contracts is a legal and cultural issue and an index of how well cultures and legal systems de facto honor the inventors will be used here.

5.2.3 *Regional and World Prices*

The prices are determined by the type of two-part tariff used in the contract. These prices may, and do, differ between geographic regions due to treatment of incomplete contracts, institutional rules, and limitations. There are therefore *regional* prices and *world* prices. It is important that the prices accounted for are categorized by region: for example, contracts between inventors of South American with the United States, Africa with Europe, Southeast Asia with China, and so on. These regional prices may be a result of the institutional policy.

In addition to negotiated prices, court do impose damages. The court settlement mechanism thus provides an important source for the economic value placed on the ideas, plus penalty. However, 95% of all court cases are settled out of court.

Many transfers do take place between companies owned by the same group, which creates “transfer prices.” These need to be separated from other prices so that they would not distort the third-party contract prices. A special survey is likely needed for this or ownership structure (but much of IP ownership is intentionally “hidden” through elaborate schemes of shell companies.)

The actual structure of holding companies is also influenced by the tax policy of nations on IP. The tax policy is therefore necessary to register with respect to certain contracts.

After transfer prices and tax optimization has been taken into account, the remaining transactions, and prices, would thus indicate the value of trade in ideas based on regional and world prices.

5.2.4 *Payments*

Payments are most commonly made using other assets than money. The principle proposed is that currency and IP assets should be used as payment assets. This means that the accounting standards need to be expanded on the asset side so that the IP right contracted will show up on the balance sheet together with the currency payments. Currency payments should include legal tenders as well as cyber cash (such as Bitcoin, Ethereum, etc.). IP payments should be accounted for by type of IP.[4]

5.3 Dynamic Economic Value Through Trade in Ideas

The proposed theory of value postulates that price = value. The willingness to pay and the willingness to accept replace the marginal cost criteria for what price should be paid for economic efficiency. This price is thus negotiated in an economic environment under rules and norms provided by the institutions.

[4] If “technical assistance,” transfer for know-how, and so on are provided as part of a contract, which is not uncommon, then these activities should be treated as separate performance obligations and accounted for as a service or Goodwill as now. Only the IP rights part should be separated out for measuring the value of trade in ideas.

5.3.1 Value from Trade in Ideas (Outcome)

The outcome is measured by the revenues from exports/imports, the international patent position (IPP), and the return on patent and other IP assets (RIPA). These three measures are further separated by technology areas, regional, and international prices.

5.3.2 Measures and Index

The outcome of the economic system can then be measured using measures based on the data collected on the economic system. New measures and index are proposed to diversify and complements current measures.

5.3.2.1 Measures

The proposed principles are necessary to create the measures of the outcome or estimate the performance of the economic system, in particular terms of trade, relative prices by tech areas, regional prices, international prices, return on IP assets, gains from specialization and other key parameters, and an international patent position. In the next section, these principles will be "translated" into statistical principles suitable to collect the data needed.

5.3.2.2 Index

To further enhance comparison between firms and nation, a new index is proposed for the level of trade in ideas. The index measures the leverage of human capital formation through the mechanism of trade in ideas (not only trade in goods or services using the IP). This index can then be used to infer the impact of institutional and taxation policy to the performance of the economic system as a whole, a policy instrument.

5.4 Data to Collect on the Economic System

The statistical system expands and restructures the existing balance of payments and financial accounting standards regarding the treatment of IP revenues, IP income, and IP assets, focusing on patents to better reflect the economic system of the twenty-first century, which will increasingly be based on trade in ideas.

The general structure of current standards is followed and expanded where possible but where the structure fails to capture the value of trade in ideas, a new structure of the data is proposed.

The same goes for data collection. Current mechanisms are not changed unless new data needs to be collected. Then new mechanisms and surveys are proposed. The goal is to provide a framework as a basis to "upgrade" and "transform" the statistics framework for the economic system of the future (see Fig. 5.1). The data results in the measuring framework for standards and trade in ideas.

5.5 Scope of Data for Statistics

The data needed for compiling the statistics on the dynamic economic system based on trade in ideas is now outlined. The actual collection of data is then discussed.

The use of the statistics is IP trade statistics and *strategic* decision-making, and the production therefore needs to meet the privacy concerns of the entities producing, trading, and using IP.

The producers of the statistics are statistical agencies summarizing data from the businesses, thus international financial accounting standards used by the companies are key to creating the statistical indicators. This makes the company (the legal entity) the "statistical unit" about which the data is collected.

To make any analysis comparative, the standards bodies need to embrace the core concepts, in particular collecting data on strategies used to reduce risk and uncertainty, which means expanding data and restructuring data and classifications in the areas of patents, patent licensing, and IP transactions (see Fig. 5.6).

5.5.1 *The Statistical Unit and Data on the Economic System*

The *statistical unit* about which transaction data is collected is the economic agent who produces, trades, and uses patents and IP. Economic agents are individual inventors, firms, and universities who have recently taken on a new economic role (Ullberg, 2015b).

Data on the characteristics of the economic system whose performance is to be analyzed is collected for the *elements* describing an economic system. The economic environment elements are (i) agents and (ii) agent characteristics, (iii) idea products, (iv) strategy, and the institutional environment consists of (v) institutions and (vi) contracts (prices and payment terms).

From the data, *measures* are created for trade in ideas.

The source of transaction data is the contract. Surveys and public data are the sources for the other elements (such as patent holdings, hierarchical structure, etc.).

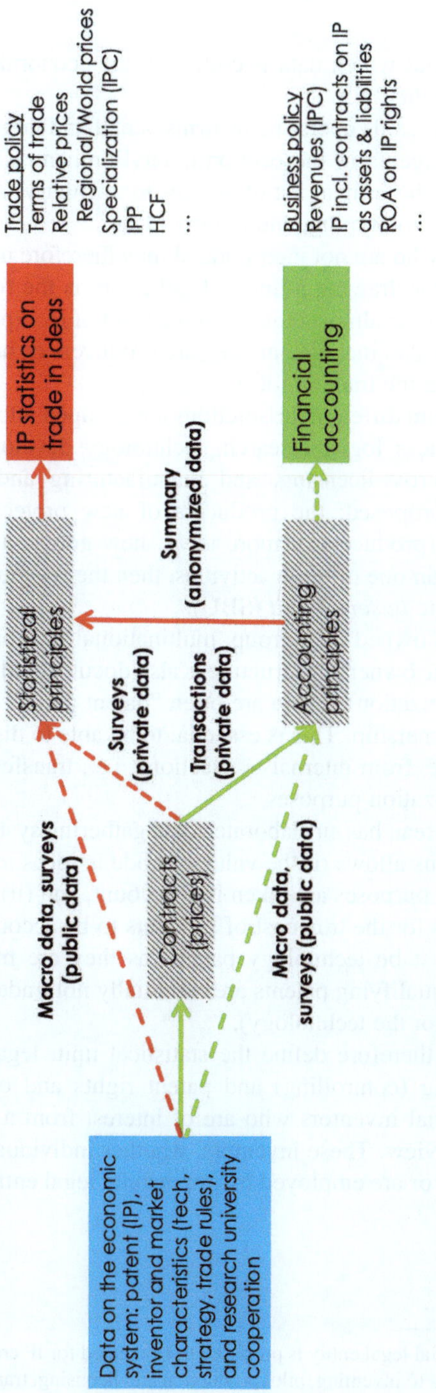

Fig. 5.6 Statistical framework components

5.5.1.1 Agents

The *statistical unit* about which data is collected, the economic agent, is the *legal entity* holding the IP rights.

The inventors/small teams operating as firms, small and large firms, universities, and institutes are the agents in the economic environment. The legal entities are classified according to the official list of entities in a country. A new legal entity is proposed for creating, prototyping, and trading ideas.[5]

Private individuals who are not incorporated may therefore not be included in the statistics. The reason for drawing a line at legal entity is the focus of cross-border trade and as such, this trade almost requires some level of incorporation. Universities that file for patents are also included as they are producers or holders of tradable IP and thereby included in the trade in ideas.

The legal entities from different jurisdictions are grouped into entities focusing on the steps of the business logic: research, technology development, prototyping, patenting, licensing, cross-licensing, and manufacturing and services. Initially, three categories are proposed: the producers of new protected technology, the traders, and the users (producers of more ideas, new goods, and new services). If an entity does more than one of these activities, then the revenue from each activity is separated by *strategic business unit* (SBU).

If the legal entity is owned by a group, multinational firm or university, holding company, and so on, the ownership structure is also documented (major work is done here by statistics organization). There are often "patent pools" and other constructs as well with shared ownership. This is essential to be able to distinguish *third-party transactions*, i.e., trade, from internal transactions, i.e., transfer prices for strategic purposes or tax optimization purposes.

The US statistics bureau has an elaborated data gathering system for this. The firm ownership structure thus allows (i) the value of trade in ideas to be calculated, once corporate and taxation purposes are taken into account, and (ii) the ownership to be clarified, which allows for the true cost of IP assets to be accounted for.

The type of IP must be technology patents as they are most tradable ("petty patents" or other less qualifying patents are essentially not tradable as they typically protect the products, not the technology).

These two criteria therefore define the statistical unit: legal entities, including their hierarchy holding (controlling) and patent rights and other tradable IP. In particular, it is the serial inventors who are of interest from a sustained economic development point of view. These inventors, whether individuals or small teams or other, typically do use or are employed by impersonal legal entities for their trade in ideas activities.

[5]The creative firm: A special legal entity is proposed to be created for IP creating and trading firms. Its line of activity is limited to inventing, pilot production and licensing, transacting in IP rights. See separate proposal, for example in (Ullberg, 2012, "Proposition 2", p.153).

5.5.1.2 Agent Characteristics

The legal entity with patent ownership/control (statistical unit) is characterized by patent ownership/access rights, human capital formation (education level of employees), scientific publications references (citing patents), and citation indexes (and other).

The ownership/access is defined as both ownership of patent and contracts on patents where the patent rights are characterized as "production factors." The patents are characterized by technology area, validation country, claimed fields of use, and remaining patent time (with extensions).

The scientific publications characterize the "quality" of the research (citation index) and whether patents or patent applications (owned by the entity or others) are cited in the publications (making visible a "transformation mechanism" between research and patents).

Research heavy entities are characterized by the publications as a measure of research projects. If project can be identified – which is often the case through research funding or presentations – it demonstrates the "tastes" for certain knowledge areas intended to deliver patented technical solutions.

5.5.1.3 Idea Products

The things traded are contracts on patent rights that constitute production assets. These things can be individual patents, "bundles" of patents, portions of patent portfolios, licenses on patents, cross-licensing of patents (with net payments), licenses on "patent pools," and a number of other combinations of patents. These trades result in a contract on the transfer or licensing of rights with payment terms, which typically include other patents, other IP, and money. The categorization of these "idea products" is needed to create meaningful statistics on the value of the trade flows.

A special categorization is proposed by *type of rights transfer* (for money, for other IP, freedom to operate, etc.) that can be related to the "bundle of rights," the production assets, that is transacted.

The proposed bundles (should) follow the practices of patent trading firms, when possible, which indicates the functioning of the economic system.

These bundles are then used (1) to establish the strategies used to reduce risk and uncertainty enabling transactions and (2) to realize their economic value. These bundles of rights include rights to use (including out-of-court settlements, court settlements, field of use, time, market) and re-trade (re-tradable contracts that can be bundled with other rights). At a minimum, transfers and licenses should be distinguished by technology area, field of use, market, and time, as well as the payment terms (money including cyber cash, or other patents and IP including cross-licensing by type of IP).

For example, there are "5G products," "renewable energy products," "clean water products," and so on where the buyer buys these production assets for their *use* (block, trade, use in goods and services).

5.5.1.4 Strategy

The management of inventions is characterized by the four strategies developed in the theory section: (1) staying clear, (2) strategic alignment, (3) marginal contracting, (4) systemic abuse.

This is the most significant data to be collected to understand the operating of the economic system.

Data on cumulative investments in patents (assets and liabilities), technology areas (by international patent classification code), markets (validation countries), time (length of rights), and fields of use (trade, goods, services) define the strategies.

- Strategy 1 is defined by a very large number of patents (typically >10,000) in a focused number of technology areas and adjacent areas, where the competition is patenting and little cross-licensing.
- Strategy 2 is defined by large number of patent (typically >1000) and also a substantial cross-licensing, giving access to more patent but from complementing technology areas.
- Strategy 3 is defined by a portfolio of patents (typically >100) and multiple transfer and licensing contracts with a range of partners.
- Strategy 4 is defined by a relatively small portfolio (typically 1–100) and heavily involved in infringement law suits as a threat where most suits are dropped in favor or out-of-court settlements.
- The data on patent assets and revenue streams are complemented by survey data.

5.5.1.5 Institutions

The institutions are characterized as the sum of the communications rights and ownership rights of entities. The purpose is to relate the strategies and economic outcome to these institutional rules. These rules are of two types: norms and formal rules. The norms are related to culture. The cultural inclination to "honor inventors" is therefore measured. The formal rules focus on the rights of entities to engage in trade given regional and international free trade agreements and how well they "honor inventors." This includes informal enforcement mechanisms (such as "shunning" misbehaving firms from the business community) and formal ones (such as penalties decided in a public court).

The institutional rules leading to honoring the inventor is the central piece in reducing risk and uncertainty in investing in and using patent rights. Trade agreements and court proceedings and business community data are needed defining how well this is done *in practice*.

This can be achieved by the creation of an index for "institutional quality" for trade in ideas, complementing current indices on enforcement.

5.5.1.6 Contract

The data on the contract defines the centerpiece of the transactions, i.e., the price and terms. However, as contracts are always incomplete, additional information on culture (and other dimensions) related to honoring the inventors are needed, just as for institutions.

The transactions emanate from the contract and are the basis for financial accounting, and, in summary, form the value of the flow of cross-border trade in ideas. The following data is collected on the idea product:

- Price – which can be a two-part tariff of a fixed fee and royalty.
- IP – patents and other IP are commonly transacted in transfers and cross-licensing contracts.
- Technology area – the technology transacted is categorized according to the international patent classification codes.
- Markets – the countries in which the patents are validated.
- Time – type of contracts can be either for a period or patent life.
- Field of use – which limitations to use, such as list of products are contracted.

The terms of payment may be based on a two-part tariff for money or for other assets or a combination. The following data is collected on the terms of payment by types of payments:

- Money
  - Fixed fee – upfront, annual
  - Royalty – based on user's revenues or other royalty-base
  - Currency, cyber cash
- IP
  - Type of IP (patents, data, trademarks, etc.)
- Equity
  - Shares
- Other
  - Work contract, time, fees (this may be considered as another performance obligation than IP)

5.6 Standards for Data Collection

The data on the economic system is collected based on accounting and statistical principles. The standards and principles proposed follow the current structure where possible. The financial accounting standard follows and expands on IFRS-15 and AIS-38, while the balance of payment statistics follows and build on BPM5 and BPM6, as well as providing input to BPM7.

The basic accounting principle for revenue recognition is, as in IFRS15, that all revenues emanate from the contract. New principles are (1) differentiation of revenues in ideas, goods, and services based on the production recourses used to produce them, and (2) that *patents (IP)* are considered production *assets* and contracts on patents (IP) are considered *assets* (for the licensor) and *liabilities* (for the licenses). The assets are seen as a "bundle of production rights." Thus, this principle is also valid for tangible assets (factories, land, capital) and intangible assets (knowledge, systems, information, relations). These two principles allow for the differentiation of revenues and the recognition of the production rights and contracts on production rights as assets. The performance of management of inventions can then be estimated dividing the returns from each product with the assets used to produce them. The value of flows from trade in ideas (IP) can be calculated as an international patent position (ownership and licensing of rights).

The data sources are therefore mainly from the *contracts* (financial accounting data), statistical *surveys* with the entity or the trade commissions, or existing *public data* sources.

The information collected on the statistical unit, the legal entity, differentiates the revenues by the assets they are created from: ideas (IP), goods, and services. Ideas are only IP, goods, and services with dominant IP content, goods, and services with some IP content. Goods and services without IP content are treated as is.

Ideas products are created from patents (IP), goods from tangible assets, and services from intangible assets (excluding IP). Ideas that do not have IP but are traded as knowledge, know-how, technical support are considered services as they are produced by intangible assets (knowledge and systems (information)).

See Tables 5.1, 5.2, and 5.3 for an overview of theory, data, standards, and definitions.

Table 5.1 Part I: Standards for data collection on the economic system, 1. Agent, 2. Agent characteristics, 3. Products traded, 4. Strategy

Economic system: twenty-first century standards for statistics and trade in: ideas, goods, and services

Element	Principles	Data collected	Practices and standards	Standard	Description
1. **Economic environment**					**Agent, agent characteristics, products, technology, knowledge, preferences, learning**
1. Agent					**Individuals/small teams, firms, universities**
	Legal entity capable of international transactions	Legal entity owning, trading, using patent (IP) rights	Statistical unit	IFRS15	
	Business unit	Strategic business unit (SBU)			Business units within firms focusing on IP trade
	Hierarchy of entities	Ownership structure			Head of ownership/control of IP assets
	Cluster of entities	Local/regional/global cluster			Business environment (knowledge context) of entity
2. Agent characteristics					
Management of inventiveness	Patenting (IP) transformation	Patenting activity	Patent citations in scientific publications	IFRS15/ New	Contribution to the stock of technology
	Patents (IP) as tradable production assets	Patent ownership by tech area, field of use, market, and time	Value of patents (IP) by cost	IFRS15/ New	Ownership of technology accounted for by cost
	Patents (IP) bundles as tradable production assets	Patent access rights by indirect ownership (pooling, other) by tech area, field of use, markets, and time	Value of patents (IP) by proportion/ share in the pool, etc.	New	Co-ownership (or other mechanism) of technology

(continued)

Table 5.1 (continued)

Economic system: twenty-first century standards for statistics and trade in: ideas, goods, and services					
Element	Principles	Data collected	Practices and standards	Standard	Description
	Contracts on patents (IP) as tradable production assets	Patent access rights (licensing) by tech area, field of use, markets, and time	Value by patent (IP) rights by price of access	New	Licensing of patented technology (IP) from others where the public or private price of the licenses
	Contracts on patents (IP) as tradable production liabilities	Patent access rights (licensing) by tech area, field of use, markets and time	Value by patent (IP) rights by price for access	New	Patent liabilities are contracts on out-licensed owned or contracted patent (IP) rights
Human capital formation	Access to human capital formation (employees)	Education level of staff	Scientific publication index	New	Contribution to the stock of knowledge
3. Products					**Commodities traded are defined by the bundle of rights used to produce them**
	Non-embodied ideas (IP)	Revenues from IP transfer contracts	Revenues from transfers	AIS38	Transfers of patent rights, bundles of rights
		Revenues from IP licensing contracts	Revenues from licensing	IFRS15	Licensing, cross-licensint
	Partly embodied ideas (IP)	Revenues from dominant IP contracts	Revenues from transfers	IFRS15	IP rights embodied in a good
		Revenues from dominant IP contracts	Revenues from performance	IFRS15	IP rights embodied in a service (in relation to a good)
	Embodied patents (IP)	Revenues from goods with some IP contracts	Revenues from transfers	AIS38	Ideas with an "add on" products
		Revenues from services with some IP contracts	Revenues from performance	IFRS15	Ideas with an "add on" services
	Goods (TA)	Revenues from goods without IP	Revenues from transfers	AIS38	Manufactured goods in global value chains
	Services (HC and systems)	Revenues from services without IP	Revenues from performance	IFRS15	Services and software as a service

Table 5.2 Part II: Standards for data collection on the economic system, 4. Strategy, 5. Institutions, 6. Contract

Economic system: twenty-first century standards for statistics and trade in: ideas, goods, and services					
Co element	Principles	Data collected	Practices and standards	Standard	Description
2. Strategy					**Strategy is the independent variable managers can change, given institutions**
4. Strategy					**Investment strategy of firms**
	Differentiation of strategies in terms of trust in each other's actions	Investments in patents (IP) and type of contracts used	Strategies	New	(1) Staying clear, (2) strategic alignment, (3) marginal transactions, (4) system abuse
	Differentiation of business models enabled by strategy	Business model in use	Business model	New	(1) Hierarchy, (2) network, (3) niche, (4) rent seeking
3. Institutions					**Institutions are the independent variable policy makers can change, given ideology, culture**
5. Institutional rules and norms					
	Institutional rules	Treatment of patent (IP) contracts in regional and international treaties	International classification of trade rules for IP	Existing/new	Formal and informal rules of trade classified from international treaties and cultural norms
6. Contract					
	Prices (international transactsions only)	Prices on contract between third parties	Two-part tariff prices in exchange for money, other patents (IP), equity or other	IFRS15/New	**Third-party transactions**

(continued)

Table 5.2 (continued)

Economic system: twenty-first century standards for statistics and trade in: ideas, goods, and services					
Co element	Principles	Data collected	Practices and standards	Standard	Description
		Prices on contracts within hierarchies with centralized or decentralized IP ownership		IFRS15/ New	Transfer prices within a hierarchy due to corporate management considerations
		Prices on contracts within hierarchies		IFRS15/ New	Strategic considerations for IP ownership (legal strengths, taxation, other)
	Incompleteness of contracts	Culture of trading nations	Sociological classification	New	Culture of honoring the inventor on a scale
	Partners countries	Country of ultimate transacting partners	Internatoinal country codes		The national ownership of patents (IP) by head of hierarchy
	Patents	Patent and patent family numbers			List of patents and patent families included in the transaction, license
	Performance (time)	Length of contract (patent life, time)		IAS38	Type of contract: patent life, time limited (guillotine)
	Payments	Money, currency, other patents (IP)	Cash, credits or barter trade income, and balance sheet	IAS38	Type of payments: money, royalty base, other patents and assets

Table 5.3 Part III: Standards for measures on the economic system, 7. Outcome measures

Economic system: twenty-first century standards for statistics and trade in: ideas, goods, and services					
Co Element	Principles	Data collected	Practices and standards	Standard	Description
Measures					
7. Outcome measures					
Economy	Terms of trade	Export/import prices of technology by technology area, field of use, markets and time		New	Wealth creating trade, export/import prices
	Relative prices	By technology area, market, field of use, time		New	Data on trade and specialization and gains from trade through relative prices
	Specialization	Technology area, markets, field of use, and time		New	Stock of patents
	IPP	Patent ownership at company level and citations in scientific papers		New	International patent position: who owns the patents in use
	HCF	Scientific publications		OECD	Stock of knowledge
	...				
Business	Revenues from IP	Revenues originating from patent (IP) assets		New	Revenues originating from patent (IP) rights whether embedded or non-embedded
	ROA	Income (return) from patent (IP) assets		New	Return on patent (IP) assets
	...				
Index	**Index of trade in ideas**	**Trade flows**		**New**	**Index of value of trade flows from trade in ideas**

Chapter 6
Description of the Collection and Compilation of Data for Each Element

The data are collected from private and public contracts, public macro data, and public and private micro data. Summarized financial accounting data and surveys are used to create national balance of payment account data and keep privacy of contract partners' data (see Fig. 5.6). First, the principles and changes to data collection are discussed, as well as what this means in terms of new *sources* of data, followed by a discussion on the measurements.

6.1 Changes in Principles and in Data Collection Structure

The general principles and standards of IFRS-15, BPM6, and other related standards that apply to the elements of the framework are used, but with the following additions to principles, structures, data, and standards. This dataset will represent "the whole" of the standards and new data needed to measure trade in ideas.

The data collection follows the principles, structural changes, and additions presented in Chap. 5. These principles are first *summarized*, and then the proposal for standards for data collection is developed, and finally, how the data is compiled into measures is summarized in table format.

6.1.1 Principles

The changes in principles include (i) a dynamic (time dependent) business logic and (ii) treating patents (IP) as assets (production rights) and contracts on patents (IP) as assets (licensee) and liabilities (licensor). Agents are limited to (iii) legal entities that can hold and trade patents (IP).

E. Ullberg, *Intellectual Property Statistics*, Contributions to Economics,
https://doi.org/10.1007/978-3-031-36386-3_6

6.1.1.1 Trade Is First of All in Ideas

The principle of trade being first of all in ideas introduces a *business logic* into the financial accounting and statistics, where new ideas – separated from other flows – are then transformed into goods and services servicing the goods. Ideas are focused on patents and other IP, or a “thing in action,” whereas goods are products such as cars, a “thing in possession.” Services include guarantees, financial, technical, performance-oriented or risk-sharing in nature “servicing” the goods. They are akin to a “thing in action” but based on human capital often supported by systems, information, and a relationship with the user of the goods and therefore have to be managed together with the users’ processes (Ullberg et al., 2002). Today’s principle of “things in possession” should be *replaced* by the principle of “things in action” as basis for what is traded.

The logic that there is an exchange in ideas to develop and select the best ideas *before* any goods or services are produced can be introduced in the statistics by separating ideas from goods and services. This separation is based on “things in action,” i.e., “bundles” of rights, which are used in the production of ideas, goods, and services. This logical separation then allows for a dynamic economic understanding of idea production:

$$\textit{exchange in ideas} \rightarrow \textit{specialization in ideas} \\ \rightarrow \textit{productivity gains in idea production}$$

The exact rules for determining which bundles of rights are associated with IP need to be determined. The proposal here is IP lifetime, i.e., the time from granting to exhaustion or nullification or otherwise taking away the rights in a country (such examples may include military use, security concerns, force majeure of which there are many cases or changes in patentability criteria, “carve outs,” moratoriums, and so on). This is the time during which the rights can be impersonally traded.

6.1.1.2 Patents (IP) are Considered Production Assets

The second principle is to consider all bundles of rights as “production asset.” As all property rights whether intellectual or other are construed as having limitations to what economic activity, they can be used for this principle works for all types of assets. This means that patents (IP), being production assets, should be capitalized on the balance sheet. Since the value of these assets is “impossible” to assert ex ante, as outlined in the theory of value, the *costs* of developing or purchasing or licensing these production assets are capitalized. They are then treated as any production asset, i.e., they are depreciated, revalued periodically, or written off based on whether they have any value for generating *future* revenues for the firm.

This means that the value generated through trade is a market value, but the assets value is a book value, not a "market value," but it allows the calculation of returns on IP assets, i.e., *income* from IP assets. The income is dependent on the price, which is dependent on the strategy, which depends on the institutions. Thus, even with a book value, relevant measures of the efficiency in the *management of inventions* and overall *economic performance* from trade in ideas can be created. The exact rules for depreciation of IP assets need to be determined in dialogue with standards bodies.

6.1.1.3 Legal Entities as Statistical Units

The principle for the statistical unit is *economic* agents that are limited to *legal entities*. These include personal companies, SMEs, MNCs, and universities/institutes, but not individuals without a legal entity. The reason for this is that international trade often requires a legal entity in practice to handle international transactions and that it is first of all the *serial inventors* that are of interest from a sustained perspective of the production of ideas and they typically manage the inventive business through legal entities (for all the reasons outlined above). The national (or regional) official registries of firms can then be used to identify the patents (IP) holders and their hierarchical structures.

This is not a necessity limitation in principle, but rather a narrowing of the scope of statistics to international accounts and balance of payments, which may make such statistics more readily accessible. There is no conceptual limitation to also include individual inventors acting as physical persons, but that may require some special treatment of data at the national level. The patent systems record inventors as individuals but the owners (assigned by the inventor) as individuals or legal entities. It is thus the *owners* of the patents (IP) that are the statistical unit, not the inventors. The inventors are critically important when it comes to their inventiveness, education (HCF), and relation to the trading firms. The nationality of the inventors is recorded in the patent system (as "national" or "foreign"), which is carried over in the statistics through the relationship with the owner.

The individual inventors or small teams of inventors are important to discover the inventiveness of a team or firm.

6.1.2 Structural Changes

Structural changes include (i) revenue recognition, (ii) recognition of assets and liabilities for financial accounting and (iii) current account, capital account for balance of payments, and (iv) a new statistical statement on an international patent position.

6.1.2.1 Revenue Recognition (IFRS15 Related)

Given that patents (IP) are considered assets, products are defined as ideas, goods, and services (there appears to be no formal definition of "product" in neither standard today!) based on the *provenance* of the revenues (which assets are used to produce them), i.e., IP, Tangible Assets (fixed assets), and Intellectual Capital that excludes IP (the human capital and systems a producer has access to).

The *revenues* are therefore classified as "ideas," "goods," and "services" in the financial accounts.

6.1.2.2 Recognition of Assets and Liabilities (IAS38 Related)

Classifying revenues according to provenance is a proposal for a standard based on an economic argument of "bundles of production rights," which has meaning conceptually and practically in both business and economics. This forms an integrative view of a firm's assets and an economy's assets.

The balance sheet therefore needs to add accounts for IP and contracts on IP as assets and contracts on IP as liabilities.

To distinguish between strategic assets (reducing uncertainty and options for future development, use, and licensing) and operational assets (reducing risk), different accounts may be needed. The strategic assets are used to determine the strategies of agents, which in turn are directly a consequence of agreed international trade rules. Establishing the strategy–trade rule relationship allows the calculation of institutional performance, which is a key input to policy.

6.1.2.3 Current Account (BPM6–7 Related)

For the balance of payment, a new *current account* is proposed for "trade in ideas" to complement "trade in goods" and "trade in services," creating a similar differentiation. The trade in ideas is further differentiated by *type* of IP, which may thus include future not yet construed rights. The idea is to use the IP defined in TRIPS, the international agreement on ("minimal") standards for granting the rights.

There may be additional accounts needed for the *capital account*, such as FDI for idea production also differentiated by type of IP invested in.

6.1.2.4 International Patent Position Statement (New)

Since IP ownership is not covered in the balance of payment manual nor the Frascati Manual (the measurement of scientific, technological, and innovation activities), a new statement is proposed for an *international patent position* (IPP). The IPP would be the equivalent of the international investment position (IIP) for financial assets.

The IPP is the accumulated *value* of a country's patent (IP) assets in other countries and its liabilities to other countries. It states who owns the patents (IP) "production assets" used in a country and thus reflects how well a country – of special importance to developing countries – leverages its human capital formation through markets in patents (IP).

6.2 New Data to Be Collected

The additions to information created are essentially related to contracts and payments, and strategy and institutions. These are micro- and macroeconomic data that are agent specific.

6.2.1 *Price Data Collected from Contracts*

The *prices variable* determines the value of the trade flows from trade in ideas. These are based on the (i) "contracts" between the trading partners and (ii) "cultural norms" dealing with incompleteness of contracts. The (iii) "payments" may be made in currency and in other IP rights, affecting other balance sheet items than liquid accounts.

6.2.1.1 Contracts

Contracts are often "linear" with a fixed fee and a royalty component. The payments may result in exchange and/or transfer of money, other IP, equity, or other assets. This is often done with a two-tier tariff: an initial lumpsum or annual fee plus a royalty on some royalty base, often revenues. It is important to distinguish the fixed fee payments from royalty, as the first represent a risk transfer and the second a risk-sharing.

6.2.1.2 Cultural Norms

All contracts are incomplete, and the way conflicts are resolved matter for the price actually paid. This is registered as an "index" of honoring the inventors. These should have agency, and not simply be institutional in character, to reveal risk-bearing and risk-sharing during the contracts time. Suitable indices need to be defined in the standards.

6.2.1.3 Payments

The payments are made in currency and other IP rights, which means that other balance sheet items than money are affected by the payments. In cross-licensing, the net payments are made in money, but huge values are made accessible through IP rights. The goal is to capture the value of these licensed rights as IP assets and liabilities.

6.2.2 Strategy

The strategies used by firms are collected by surveys or deduced from the patent (IP) assets and liabilities, and types of contracts used to generate revenues. The proposed four strategies will be used as standard (see Chap. 2).

6.2.3 Institutions

The institutional *rules* used in poly-lateral and multilateral free trade agreements need to be characterized and classified to provide a standardized way to measure institutional rules. These rules are the *independent variable* in any assessment of which strategy is used and thus which level of cooperation is incentivized by the rules. The institutions are thus a statistical unit.

6.2.4 Summary Table of Data Collection

See Table 6.1 for the *principles* on which data shall be collected on the proposed framework. See Tables 6.2 and 6.3 for the *data structure* of the proposed framework.

6.2.5 Statistical and Financial Accounts

See Tables 6.4 and 6.5 for the current and proposed *accounts* to be used for accumulating the data in the proposed framework. Current account names and codes are used for reference (see Chap. 5).

Table 6.1 Principles on which data shall be collected

Economic system		Statistical system			
Idea based for the twenty-first century		Business policy: Management of Inventions Financial Performance Statistic		Trade policy: Balance of Payment Statistics on trade in ideas	
Principles	Comment	IFRS15/ IAS38	Proposal	BPM5,6 and IIP	Proposal
Revenues and income (i)					
Are trade outcomes a result of static or dynamic exchange?	The outcomes are decided by output/input, where the output is decided in a competitive market in patent (IP) given agents' strategies and trade rules and taxes	Static outcomes over a period	**A dynamic outcome (learning,) given inventor strategy and institutional rules and taxes, where trade is first of all in ideas**	Static outcomes over a period	**A dynamic outcome given inventor strategy and institutional rules and taxes, where trade is first of all in ideas**
Where do revenues originate from?	The source of all revenues are the contract	The contract	**The contract (no change)**	NA	**NA**
Where to payments originate from?	The revenues are separated from payments in the standards. Revenues data from contracts and, BoP data from payments	The contract	**The contract (no change)**	Cross-border economic transactions flowing from contracts	**Cross-border economic transactions flowing from contracts (no change)**
What do payments consist of?	IP transactions may result in the exchange of IP and other assets, not only money	Money, other financial assets	**Money, IP, equity, and other financial and non-financial production assets**	Money	**Money, but IP and other assets are accounted for in the IPP**
What is traded (contracted) in the economic system?	The economic system is seen as a set of independent but interacting processes, where trade in ideas is first, then goods, then servicing the goods	"Goods and services" (not a defined accounting concept)	**Ideas, goods, and services (defined by provenance)**	Trade in goods, and in services (each account reflects a single economic process. I. B.2)	**Trade in ideas, goods, and in services (each account reflects a single economic process based on a business logic)**

(continued)

Table 6.1 (continued)

Economic system		Statistical system			
Idea based for the twenty-first century		Business policy: Management of Inventions Financial Performance Statistic		Trade policy: Balance of Payment Statistics on trade in ideas	
Principles	Comment	IFRS15/ IAS38	Proposal	BPM5,6 and IIP	Proposal
How is cost for IP considered?	The valuation of IP is done by the cost of developing, maintaining, devending, or purchasing the rights which is capitalized	Market value	**Book value**	NA	**NA**
How is depreciation of IP assets considered?	The accumulated cost are capitalized and depreciated	As goodwill	**Annual revaluation or write-off of the future value of the IP assets with respect to potential future revenues**	NA	**NA**
Assets and liabilities (ii)					
What is intellectual property, IP?	IP is considered a bundle of production rights defined as property rights on things in action, an economic asset	Intellectual property (not defined)	**A bundle of rights related to intellectual property as defined by property rights on things in action, such as TRIPS, other IP right systems**	"Proprietary rights not included elsewhere"	**A bundle of rights related to intellectual property as defined by property rights on things in action, such as TRIPS, other IP right systems**
What are contracts on IP?	Contracts on IP assets are considered production rights, assets/ liabilities	Short term financial obligations	**Long-term production assets and liabilities**	NA	**Claims and obligations accounted for in the IPP**

(continued)

Table 6.1 (continued)

Economic system		Statistical system			
Idea based for the twenty-first century		Business policy: Management of Inventions Financial Performance Statistic		Trade policy: Balance of Payment Statistics on trade in ideas	
Principles	Comment	IFRS15/ IAS38	Proposal	BPM5,6 and IIP	Proposal
Where do IP licensing show up?	IP licensing (idea exchange) is considered an own economic process, an exchange in access to production rights (assets)	A service	**An idea**	Charges for the use of IP	**Trade in ideas (a proposed new current account)**
What is IP transfer?	IP transfers (idea exchange) is considered an own economic process, an exchange in production rights (assets)	A sale of assets for goodwill	**A sale of a production asset for money, other IP or other assets**	Other business services	**Trade in ideas (a proposed new current account)**
Statistical units (iii)					
What is the statistical unit about which data is collected?	There are two statistical units: the developing country inventor's formalized as the type of legal entity (often individuals/small teams and SMEs) *and* the institutions, frameworks, and international cooperation deciding the trade rule, in turn, creating the incentives for the agents' strategies	The legal entity	**The legal entity of the agent *and* the institutions, frameworks or international cooperations establishing treaties regarding the trade rules**	The legal entity	**The legal entity of the agent *and* the trade rules or institution (could be a bilateral, poly-lateral or multi-lateral treaty)**

Table 6.2 Data to be collected on the economic environment and strategy

	Economic system		Statistical system			
	Idea based for the twenty-first century		Business policy: Management of Inventions Financial Performance Statistics		Trade policy: Balance of Payment Statistics on trade in ideas	
	Micro data (metadata)	Source	IFRS15/IAS38	IP Trade Statistics	BPM5,6 and IIP	IP Trade Statistics
1.	**Agent data**					
	Legal entity	Public registry; legal type	Entity	Negotiated public/private price	Anonymized entity data	Summary by entity type
	Hierarchy	Surveys/public registry	(Direct ownership > x%)	Transfer price within hierarchy	(US BEA and EU project)	″
	Tax holding entity	Surveys/tax optimization; transfer prices		Transfer price		″
	IP holding entity	Surveys/IP management; transfer prices		Transfer price		″
	SBU	Surveys/strategic business unit		Transfer price		″
	Network entity	Surveys/patent pool, other network		Negotiated transfer price within network		″
	Third-party entity	Third-party transactions; trade in ideas		Negotiated public/private price with third party	(US BEA)	Trade in ideas
2.	**Agent characteristics data**					
	Business model	Surveys/business model of agent or strategic business unit within agent		Central business activity by strategic focus		Economic process, type
	Technology	Technology areas (IPC)		IPC classification		IPC classification
	IP rights	Value	Goodwill	Cost		Access to IP rights

	Patent rights	Contracts/WIPO*, patent offices/special services; courts		IPC; patent legal status		Access to patent rights
	Other IP rights	Contracts/IP registry or non-registered IP		Other IP right by type of rights		Access to other IP rights
	Open access technologies	Exhausted IP/non protected technology		IPC classification		IPC classification
	Knowledge/ research					
	Human capital formation	Company surveys		OECD codes		OECD codes
	Patent referenced in academic publications	Private/public databases				
	Search systems	Company surveys		Elsevier, others		Summary of entity data
	Research budget	Financial accounts	R&D budget	Research areas		"
3.	**Product data**					
	Ideas, goods, services	Annual statements	Revenues from goods and services	Revenues from IP, TA and IC w/o IP rights	Export/import of goods/ services	Export/import of ideas, goods, services
4.	**Strategies data**					
	Strategies pursued (used) by agents	Surveys/financial statements; type (stay clear, align, margin, abuse)		**Strategy by type**		Summary of entity data
	Patent portfolios	Patent portfolio size, tech area, contracts and court cases		Patents by tech area, validation countries, time		"

(continued)

Table 6.2 (continued)

	Economic system		Statistical system			
	Idea based for the twenty-first century		Business policy: Management of Inventions Financial Performance Statistics		Trade policy: Balance of Payment Statistics on trade in ideas	
	Micro data (metadata)	Source	IFRS15/IAS38	IP Trade Statistics	BPM5,6 and IIP	IP Trade Statistics
	Contracts	Patent portfolio size, tech area and contract type		Patents by tech area, validation countries, time		″
	Research strategies based on portfolios	Tech area direction of research by research area, future patent portfolio size by tech area		OCED science areas, IPC classifications		″

Table 6.3 Data to be collected on institutions and contracts, and summarized data

	Economic system		Statistical system			
	Idea based for the twenty-first century		Business policy: Management of Inventions Financial Performance Statistics		Trade policy: Balance of Payment Statistics on trade in ideas	
	Micro data (metadata)	Source	IFRS15/ IAS38	IP Trade Statistics	BPM5,6 and IIP	IP Trade Statistics
5.	**Institutions data**					
	The institution	Institutions, frameworks and international cooperations negotiating free trade agreements		**Type of institutional rules in use in trade**		**Type of institutional rules**
	Trade rules	The free trade agreements; categorization of rule		Agreement reference		Agreement reference
	Property rights on communication	Patent laws and trade rules which structure exchange (trade)		Trade rule reference		Trade rule reference
	Property rights on things in action	Patent laws (TRIPS) which define the rights and enforcement		Property right rule reference		Property right rule reference
6.	**Contracts data**					
	Linear contracts	The price paid for the IP	Revenues from services	Revenues from IP rights	Trade in services	Trade in ideas
	Fixe price	Fixed price (initial or annual)		Fixed price revenues (risk bearing)		Sum of fixed price exp/imp
	Royalty price	Royalty price (based on sales, other)		Royalty revenues (risk sharing)		Sum of royalty exp/imp
	Court settlement	Official penalty to receive/pay	Recognition of liability	Court settlement		Sum of court payments
	Out of court settlement	Public/private settlement to receive/pay	Recognition of liability	Out of court settlement		Sum of out of court payments
	Parties	Agent ID and country codes of parties	Entity	Legal entity and type	Countries trading	Countries trading

(continued)

Table 6.3 (continued)

Economic system		Statistical system			
Idea based for the twenty-first century		Business policy: Management of Inventions Financial Performance Statistics		Trade policy: Balance of Payment Statistics on trade in ideas	
Micro data (metadata)	Source	IFRS15/ IAS38	IP Trade Statistics	BPM5,6 and IIP	IP Trade Statistics
Technology (patents)	Patents, other IP		IPC; patent legal status		Access by ownership or contracts
Incompleteness of contracts	Cultural norms rules used to settle differences		Resulting outcome in prices		Resulting outcome on prices
Summarized data	**Source**	IFRS15/ IAS38	IP trade statistics	BPM5,6 and IIP	IP Trade Statistics
Contracts	Contract			Countries trading	Countries trading
Prices	Contract			Goods and services prices	Regional and world prices
Strategies	Surveys/financial statements				Strategies in use
Institutions	Institutions, frameworks, and international cooperations negotiating free trade agreements				Institutional rules in use
Macro data	**Source**	IFRS15/ IAS38	IP Trade Statistics	BPM5,6 and IIP	IP Trade Statistics
Patent data	Patent offices, agent data				Country patent ownership by tech area
HCF	Agent data				Country HCF data
Articles w. Ref to patents	Academic research data, agent data				Integration sci&tech
Investments in research	Academic research funding data, agent data				Investment OECD research area type

Table 6.4 Income statement: Current and proposed accounts for standards for revenues and cost

Economic system		Statistical system			
Idea based for the twenty-first century		Business policy: Management of Inventions Financial Performance Statistics		Trade policy: Balance of Payment Statistics on trade in ideas	
Statements	Source	IFRS15/ IAS38	IP Trade Statistics	BPM5,6 and IIP	IP Trade Statistics
Sales/ revenues from products (i)	Source				Current account (iii)
Revenues from ideas (IP)	Contract	*IFRS15.1–2 and 15.26(i)*	Revenues from IP rights	Royalties and license fees and services, incl. sale of proprietary rights	Trade in ideas (IP)
Licensing of IP rights	Contract	*IFRS15. B52–53*	Licensing, cross-licensing, compulsory licensing	8.1 (franchises and trade-marks), 8.2 (outcomes of R&D), 8.3 (software), 8.4 (audio-visual)	8.1.1–2 (8.1 differentiated) and 8.2.1.1–8 (8.2 differenti-ated by IP type) each further differentiated by process (negotiation), 8.3, 8.4
Fixed cost	Contract		Fixed fee (blocking value)		Diff. of licens-ing by tariff type (risk)
Royalty cost	Contract		Royalty fee (use value)		Diff. of licens-ing by tariff type (risk)
Enforcement	Contract	*IFRS15*	Out-of-court and court settlement		Differentiation of licensing by IP type and process of negotiation (out-of-court, court)
Sale	Contract	*IAS38*	Transfers	10.1.1.2 (sale or rights from R&D)	10.1.1.2.1.1–8 (diff. by IP type)
Exchange of IP rights w. net payment	Contract		Exchange with net pay-ment (see assets)		8.1.1–2, 8.2.1.1–8, 8.3, 8.4 as above
Revenues from goods	*Contract*	*IFRS15.B54*	*Revenues from TA*	*Trade in goods*	Trade in goods (embodied IP)

(continued)

Table 6.4 (continued)

Economic system		Statistical system			
Idea based for the twenty-first century		Business policy: Management of Inventions Financial Performance Statistics		Trade policy: Balance of Payment Statistics on trade in ideas	
Statements	Source	IFRS15/ IAS38	IP Trade Statistics	BPM5,6 and IIP	IP Trade Statistics
Sales/ revenues from products (i)	Source				Current account (iii)
			rights w embodied IP		
	Contract		*Revenues from goods w/o IP*		Trade in goods (excl. IP)
Revenues from services	*Contract*	*IFRS15.B54*	*Revenues from IC rights w embodied IP*	*Trade in services*	Trade in services (embodied IP)
	Contract		*Revenues from IC rights w/o IP rights*		*Trade in services (excl. IP)*
Cost					
Cost of using IP rights	Contract	*IFRS15*	Cost of using IP	8.1, 8.2, 8.3, 8.4 as above	8.1.1–2, 8.2.1.1–8, 8.3, 8.4 as above
Fixed cost	Contract		Blocking value		Diff. of licensing by tariff type (risk)
Royalty cost	Contract		Use value		Diff. of licensing by tariff type (risk)
Research phase	Contract	*IFRS15*	(same as IFRS)	10.1.1.1 (R&D services)	10.1.1.1 (diff. by OECD research area)
Development phase	Contract	IAS 38.72–73	IP development and protection is capitalized	10.1.1.1 (R&D services)	10.1.1.1 (diff. by OECD research area)
Maintenance phase	Contract	*IFRS15*	(same as IFRS)	10.1.1.1 (R&D services)	10.1.1.1 (diff. by OECD research area)

(continued)

Table 6.4 (continued)

Economic system		Statistical system			
Idea based for the twenty-first century		Business policy: Management of Inventions Financial Performance Statistics		Trade policy: Balance of Payment Statistics on trade in ideas	
Statements	Source	IFRS15/ IAS38	IP Trade Statistics	BPM5,6 and IIP	IP Trade Statistics
Sales/ revenues from products (i)	Source				Current account (iii)
EBITDA	Financial statement				
Depreciation	Financial statement	IAS 38.74 (goodwill)	Length of economic, patent, or contract life		Depreciation 3–5y according to (new) standards
EBIT	Financial statement				
Finance	Financial statement	IFRS15	Financial cost of accessing IP asset		Financial charges for access to IP
EBT	Financial statement				
Taxes	Financial statement	IFRS15	Taxes on IP holdings		Tax paid on IP assets
EAT	Financial statement				
Dividends	Financial statement	IFRS15	Dividends related to IP holdings		Dividends from IP assets
Retained earnings	Retained earnings from investment in IP assets				

6.3 Summary of Proposed New Measures

The proposed new measures are structured in three areas, which are given as follows:

(i) the *economic system performance*, i.e., the contribution of trade in ideas to a country's economy as a whole; the leverage of their human capital formation through markets in patents (IP),

(ii) the *integration of science and technology* through markets in patents (IP), and

(iii) the *management of inventions* to create the patented technology and other IP rights and trade them in markets in patents (IP).

Table 6.5 Balance sheet: Current and proposed accounts for standards for assets and liabilities

Economic system		Statistical system			
Idea based for the twenty-first century		Business policy: Management of Inventions Financial Performance Statistics		Trade policy: Balance of Payment Statistics on trade in ideas	
Statements	Source	IFRS15/ IAS38	IP Trade Statistics	BPM5,6 and IIP	IP Trade Statistics
Assets and liabilities (ii)		Balance sheet		IIP	International patent position
Non-current assets (long-term assets)		IAS 38.12			
Capitalized development expenses	Development phase is capitalized	IAS 38.57–58			
Goodwill	Acquisition of IP	IAS 38.25–32/ IFRS 3.18–19	(same as IAS 38)		
Intangible assets	Recognition, exchange, and disclosure	IAS 38.18, 21 (recognition), 38.45–47 (exchange), 38.118–128, (disclosure)	IP are considered a production asset rights		IPP
Cost model	An intangible asset is carried at cost less any accumulated amortization and any accumulated impairment losses	IAS 38.74	All cost in developing the IP assets are capitalized, amortized and written off		IPP
Revaluation model	An intangible asset is carried at its fair value (i.e., revalued amount) less any accumulated amortization and any accumulated impairment losses	IAS 38.72–73	(IP only carried at cost, as fair value is "impossible" to meaningfully assess)		
Contracts on IP	Contracted IP asset for a limited period or patent life	(currently not part of standard)	Contracts are revaluated annually with respect to future business value		IPP

(continued)

Table 6.5 (continued)

Economic system		Statistical system			
Idea based for the twenty-first century		Business policy: Management of Inventions Financial Performance Statistics		Trade policy: Balance of Payment Statistics on trade in ideas	
Statements	Source	IFRS15/ IAS38	IP Trade Statistics	BPM5,6 and IIP	IP Trade Statistics
Assets and liabilities (ii)		Balance sheet		IIP	International patent position
Financial holdings	Equity holdings as payments for IP		Financial assets in exchange for IP		
Current assets (short-term assets)					
Contract assets	Remaining use-right to patents that are not yet invoiced		Remaining use-rights to IP that are not yet invoiced		IPP
Liabilities					
Out-contracted IP	Contracted IP licenses based on own IP or contracts on IP		The performance obligation promised to licensees by licensor		IPP
Debt	Debt funded IP development investments		Financial debt for investments in IP	(IIP)	
Current liabilities					
Out-contracted IP	Pre-payments for use of IP, i.e., annual or one-time fixed fee		Pre-payments for use of IP, i.e., annual or one-time fixed fee		IPP
Equity					

See Chap. 1 for the definitions of the measures and Table 6.6 for the summation of the financial and economic performance measures proposed in the framework.

6.4 A Trade in Ideas Index

To rank countries' performance, an *agent-based* index shall be developed. Such an index should be sensitive to key economic development technologies: food/water/shelter, energy, and ICT. The international patent position (IPP) is a key part of such an index. The development of this index is proposed to be done in cooperation with especially developing nations desiring to develop their trade in ideas.

Table 6.6 Measures for institutional and management analysis and ranking

	Economic system		Statistical system			
	Idea based for the twenty-first century		Business policy: Management of Inventions Financial Performance Statistics		Trade policy: Balance of Payment Statistics on trade in ideas	
	Measures	Calculations	IFRS15/ IAS38	IP Trade Statistics	BPM5,6 and IIP	IP Trade Statistics
7.	**Measures**					
	Regional prices	Revenues between firms in economic cooperation areas		x		x
	World prices	Revenues between firms worldwide		x		x
	Terms of trade	Export/import prices, by tech.area		x		x
	Specialization	Technology focus of a firm and country		x		x
	IPP	Patent ownership, licensed rights and obligations		x		x
	HCF	Education levels in OECD codes		x		x
	HCF focus	Top HCF areas in OECD codes		x		x
	Gains from trade and specialization (relative prices)	Relative prices of different tech areas of patented (IP) rights		N/A		x
	Integration of science and technology	Revenues from patents (IP) in areas where research is strong		N/A		x
	Economic performance	**Revenues from IP**		x	BPM5,6: 8, 10	x
	Given strategy	Revenues from IP given strategy in use for firms, country		x		x
	Given strategy, given institutions	Revenues from IP given strategy and institutional rules for firms, ECA, world		x		x

(continued)

Table 6.6 (continued)

Economic system		Statistical system			
Idea based for the twenty-first century		Business policy: Management of Inventions Financial Performance Statistics		Trade policy: Balance of Payment Statistics on trade in ideas	
Measures	Calculations	IFRS15/ IAS38	IP Trade Statistics	BPM5,6 and IIP	IP Trade Statistics
Management of inventions					x
Revenues from IP	Revenues from IP in terms		x		x
RIPA	Revenues from IP/IP assets (rights, contracts) by tech area		x		x
Strategy position: goal, transition	Strategies in use (1–4) by firms, countries, ECA, world		x		x
International patent position	Owned IP assets – licensed IP assets		N/A		Access to/ownership of key technologies for economic development
Index					x
Prices by tech area			x		Export/import prices for key technologies for economic development
1. Food/water (housing)					x
2. Energy					x
3. ICT					x
(4. Health)					(x)

Part IV
A Proof of Concept

Chapter 7
A Proof-of-Concept Study in Perú

This chapter presents key finding from a proof-of-concept study in Perú of the proposed statistical framework. It was the inventor agents that were the source of the data. The international trade rules were not discussed. That will be a subject for another study. It is a first study of whether key elements of data collection can be gathered and compiled for statistics and financial accounting of cross-border trade in ideas.

This study would not have been possible without the extensive support of a team of patent and statistics experts from INDECOPI, the Peruvian IP and consumer protection agency.

7.1 A Developing Country Study in Perú

Can the data needed to calculate key measurements on the economic system performance be collected in a developing country? If this can be done, data can likely be collected to understand international trade patterns and inform international business and trade policy.

The focus of developing country is essential for trade in ideas as 7 billion of the nearly 8 billion people live there and the human capital formation is growing rapidly in these nations. There are currently 100s of millions of highly educated people in the developing world, which was not the case only two decades ago. In less than a decade, the south (including China) will surpass the north in human capital formation, when measured in terms of academic publications.

The contribution to the stock of knowledge and the stock of technology can then be calculated. The revenues from trade in ideas, terms of trade, Return on Assets (ROA), and so on, as well as giving data on North–South trade in ideas to inform policies on trade rules.

This section reports key findings from a proof-of-concept of the proposed statistical framework based on a survey of 100 developing country inventors from Perú. The study was done with generous team-support and cooperation with INDECOPI.

E. Ullberg, *Intellectual Property Statistics*, Contributions to Economics,
https://doi.org/10.1007/978-3-031-36386-3_7

The reported findings indicate that it looks promising that data could be collected from a developing country, and the proposed IP statistics framework can therefore be used to calculate measures to inform business, university–institute, and government policy on trade in ideas. The next step would be to do a full-scale data collection for a limited number of technologies and markets.

First, the design of the proof of concept (POC) is outlined and how it was carried out. Then some key results are reported and analyzed. The section ends with conclusions on what may need to change to adopt the framework.

7.2 The Design in Cooperation with INDECOPI

The design of the proof-of-concept of the proposed IP statistics framework was done to do a first evaluation on whether the data needed could be collected and complied to national statistics in a developing country. Since the purpose of the trade in ideas program is to create incentives for North–South exchange in patented technology, it is important that quality data can be collected from the statistics institutions of smaller means and larger means.

The presumption was that the market in ideas operates in *the same way* everywhere, albeit at different maturity. A survey was thus developed along the lines of the same structure as the proposed IP statistics framework and sent out to inventors who were already patent holders.

The POC was done in close collaboration and supported – not without enthusiasm – by INDECOPI, the Peruvian IP, and consumer protection agency. A small team was created by INDECOPI to support the gathering of inventor contact information, translation of the survey in Spanish, sending out a cover letter to the survey to introduce its purpose, as well as reminders, personal contacts by phone, text messaging, and email to interested inventors to participate and complete the survey. In the end, data collected from 100 Peruvian inventors/teams/companies/universities. This number was on the high end of expectations and provides a unique dataset.

The data was then cleaned up and analyzed along the important elements proposed in the statistics and accounting framework. Key results from the survey are reported.[1]

7.3 Selection of Inventor Firms

To make a selection of inventors as representatives, a database with more than 300,000 contacts was used to reach out to inventors and then ask them to self-select to participate. It was based on the contact list that INDECOPI uses for information to the inventor community in Perú. It includes inventors, patent holders,

[1] A full report of the survey is planned to be developed at a later stage. It will likely be published in a "policy brief" or research paper covering the most interesting results. Interest in further analyzing this data can be directed directly to the author.

and anyone interested in the work of INDECOPI. INDECOPI is the public office of all IP, and the consumer protection agency of Perú. A simple question was asked whether the respondents owned patents, what formal education level they had, what institution they represented (firm, university, institute), and whether they were interested in participating in a survey on trade in ideas (IP). The survey was then sent out to these self-selected respondents.

7.4 The Survey Structure

7.4.1 *The Operation of the Economic System in a Developing Country*

The same operation of the economic system is assumed in developed and developing countries at the level of principle that is described in the model. This allows for comparable trade measurements to be calculated for the North–South exchange in ideas. This is based on the fact that the operating of the *market in patents* is the same, albeit in its infancy in developing countries and more developed in developed countries. This is a conclusion of the pilot study report on seven developing countries (2017). The differences lie in volumes, breadth, depth, height of technology, maturity of institutions, and so on.

7.4.2 *The Survey*

Based on the assumption of the same economic system, but varied use of the IP system, the conclusion is that the world economic system is highly integrated (which is a well-known fact but challenged since 2008 financial crises) but operates either with or without IP. Compare the nineteenth- and twentieth-century "industrial-based" economic system with the twenty-first-century "idea-based" economic system discussion in Chap. 1.

The data collected was from the inventors who followed the structure of the framework: (i) Economic environment, covering agent, agent characteristics, especially patents and IP, and strategies; and (ii) Institutions, covering contract types (not international trade rules). Question on the way data was accounted for in the financial accounts was then asked.

The questions were divided into three sections:

Section 1: Information on the inventor (Q1–3)
Section 2: Information on the use of patents, licenses, and other relevant IP rights including strategies used (Q4–12)
Section 3: Information on the financial accounting of respective revenues, cost, profits, finance, and taxes (Q13–23)

A summary of the questions asked in an income statement (IS) and balance sheet (BS) is shown in Tables 7.1 and 7.2.

Table 7.1 Summary of questions on revenues from patents/IP that would go into the IS and BS of inventor firms

1	Revenues		IP rights			How do you use your IP							
						International				National			
1a	General info	**IP**	Y/N	How many rights do you hold	Application ID (if possible)	Licensing	Cross-licensing	Selling	Own use	Licensing	Cross-licensing	Selling	Own use
		Patents											
		Utility patents											
		Trade secrets											
		Trademarks											
		Industrial designs											
		IC patterns											
		Data											
		Geographical indications											
1b	Patent	Patent family (market)		Validation countries									
		Field of use		Create list									
		Tech areas		IPC									
		Patent system used		PCT;EP;PE									
1c	Contracts	Contract		Prices		Work (tech assistance)	Information assets	Cross-license for technology	**Fixed fee**	**Royalty**	Other	Licence period, years	Guillotin license o patent lif
		Price when licensing (contract terms)		Patents licensed in exchange for. . .									
		Price when selling rights		Patents sold in exchange for. . .									
		Price (selling company)		M&A transactions									
		Markets		Countries licensed/ sold to									
		Counterpart		Legal entity: Univ., firm, indiv., instit.									
		Counterpart relation		Subsidiary/group entity (transfer price)									
		Cross-border/ national		Legal Inc. of seller									

										Business strategy			
fensive				Goal in 3–5years						Strategy			
ocking	Options	Publication (destroy prior art; freedom to act)	Academic publication (create prior art)	Number of patent appl.	Growth in pat. Apl. %	Patents	Growth in patent portfolio	Licensing	Growth in licensing revenues	1. Stay cllear (own use)	2. Strategic alignment/capture period	3. Simple in/out licensing or sale (marginal contract)	4. To sue for money as a business (systemic abuse)
ditional wledge IP													

Table 7.2 Questions relating to cost that would go into the IS and BS of inventor firms

2	**Cost**	
2a		*Market in patents*
		Inventing (applied research, prototyping, test)
		Patenting (application, patent, renewal)
		Licensing, selling
		Enforcement
		Patent portfolio management by tech area
2b		*R&D costs*
		Research area
3	**Depreciation**	
		Market in patents (from assets)
		R&D (from assets)
4	**Financing**	
4a	**Financial items**	
		Grants toward markets in patents
		Grants toward R&D
		Interest bearing finance
4b	**Tax breaks**	
		Tax rebate on markets in patents
		Tax rebate on research
4c	**Equity funding**	
		Capital contribution for equity shares
5	**Assets**	
		By research area
		Markets in patents investments (patenting, licensing, investment)
		Depreciation with the economic life = legal life (max 20y + 10y)
		Contracts on patents (in-licensing)
		Depreciation with the economic life = legal life of contract (3–5 years)
		R&D investments by research area
		Depreciation with the economic life of tech are patents (5 years)
6	**Liabilities**	
		Contract obligations in market in patents (out-licensing)
7	**Financial accounting**	
		Ownership relations
		Shell company
		Financial accounting standard used
		Financial accounting system used
		Audited accounts
		Public accounts

7.5 The Results

Key results of the *economic environment* cover the agent and agent characteristics and their strategies. Key results on the *institutions* cover the contract types and patent systems used. Key results on the handling of *financial information* cover financial accounting standards used for revenue recognition, cost, finance, and taxes.

7.5.1 The Economic Environment

The economic environment consists of agent and agent characteristics, as well as "IP products" sold and are strategies agents use to trade ideas. A selection of survey questions is reported to begin to answer these questions. The elements covered are management of inventions, legal entities, access to human capital formation, patent portfolio growth, finance (under accounting principles), and strategies.

7.5.1.1 The Inventor's Economic Environment

Inventions are typically made by individuals or small teams. In the case of Peruvian national inventors (inventors with permanent residence in Perú as in the international coding of patents), more than half of the inventors work in small teams and the rest are individuals. About one-third have direct access to research, working within universities. There is thus an important link with universities which then allows for the use of patents to integrate science and technology. Data on this important link can thus be captured as well as the economic organization of teams (see Fig. 7.1).

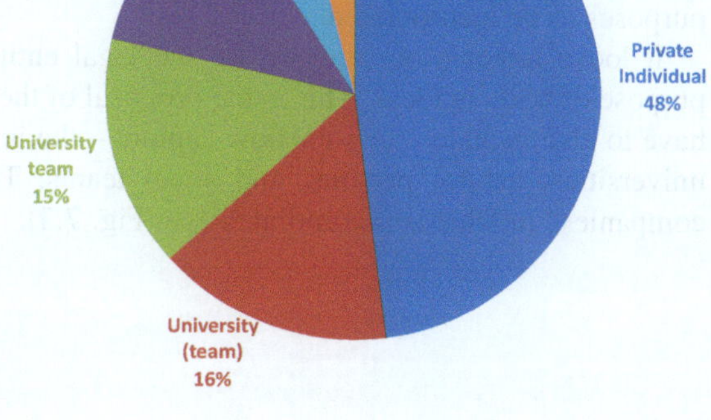

Fig. 7.1 Inventors' economic environment

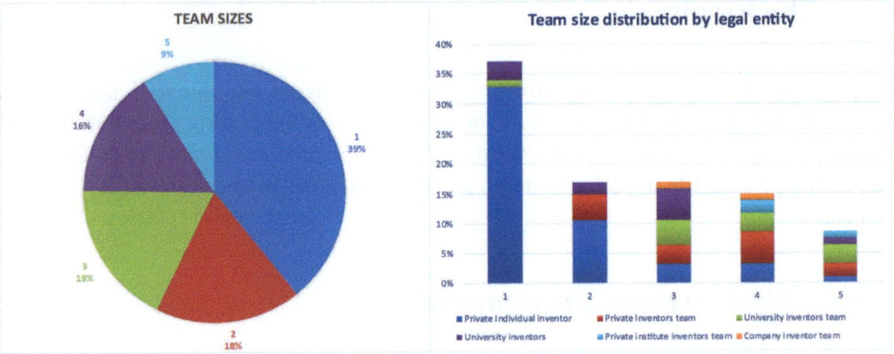

Fig. 7.2 Team sizes and legal entities

7.5.1.2 Management of Inventions

In a further investigation into the actual operating of the *management of inventions*, the size of the teams is important. Inventors are mainly organized in small teams, making up 63% of the inventors, 37% are individuals, 2–4 members make up 49%, and >4 members make up 14%. More than 50% are thus informal small teams. Legal entities typically organized their inventive activities in teams of 3–5 people. This confirms the assumption that the future management skills include making every talent productive is a small-team management issue (see Chap. 1).

This makes it clear that any statistical approach may need to be "accessible" down to the individual or at least small team level to capture the value of trade in ideas. This is a real challenge and requires special software tools for accounting and reporting (see Fig. 7.2).

7.5.1.3 The Legal Entity Used to Hold Patents

The type of legal entity used for patent ownership is dominated by natural persons or own company, owned by the individual or small team inventors. Second category are the universities. What is noteworthy is the low frequency of private Peruvian companies. In order to trade internationally, one needs for all practical and tax purposes to be incorporated.

It looks rather promising to use the legal entity as the statistical unit for the purpose of trade in ideas. This is the proposal of the framework. However, one may have to also include – or somehow capture – the informal teams that exist between universities, natural persons, and small teams. This means including "personal companies" in what is incorporated (see Fig. 7.3).

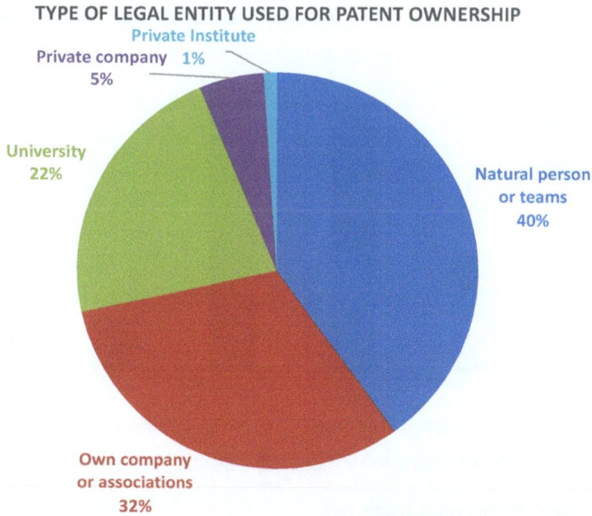

Fig. 7.3 Legal entities

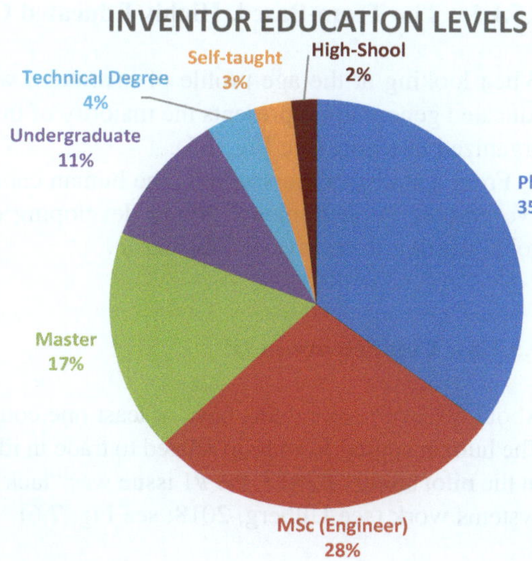

Fig. 7.4 Inventor education levels

7.5.1.4 The Human Capital Formation

Key resources include the level of education the teams have access to, which reveals the level of technology the inventors can contribute with to increase the global stock of technology. A striking result is that 80% of the inventors in Perú has either a PhD, Master, or Engineering degree; 15% have undergraduate or technical decrees; and only 5% are self-taught or have a high-school degree. This means that the rise in human capital formation in the developing countries during the last decades discussed in the introduction is present (see Fig. 7.4).

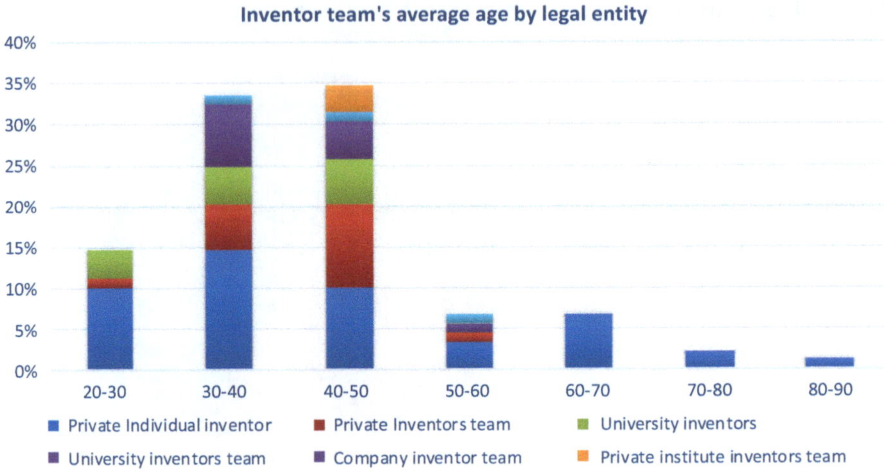

Fig. 7.5 Inventor teams age

7.5.1.5 The Team-Based, Highly Educated Generation

When looking at the age-profile of inventors, we clearly see that this new, highly educated generation represents the majority of inventors and that these inventors are organized in teams (see Fig. 7.5).

From a statistical perspective, the human capital formation present in the inventive process is phenomenal. Many developing countries may thus sit on "human gold" waiting to be invested further in.

7.5.1.6 Competence in IP

About 60% of respondents have at least one course in IP, 40% have no IP training. The human capital formation related to trade in ideas is therefore a serious limitation. In the pilot study of 2017, the #1 issue was "lack of awareness" of how the patent/IP systems work (see Ullberg, 2018; see Fig. 7.6).

7.5.1.7 The Patent Portfolio Growth

The average growth of the patent portfolio over the past 15 years is about 250 patents for 34 respondents or 1 patent/2 years. The highest frequency is for companies with 1–6 patents or about 1 patent/3 years.

These data indicate the growth of the stock of technology/company, thus how productive inventor firms are in creating tradable patents, given their investment in human capital. It also indicates the critical "serial inventors," which appear to be about three-fourths of the applicants (this number needs further study of patenting behavior) (see Fig. 7.7).

This is a measure of the management of invention. Such data are thus possible to collect.

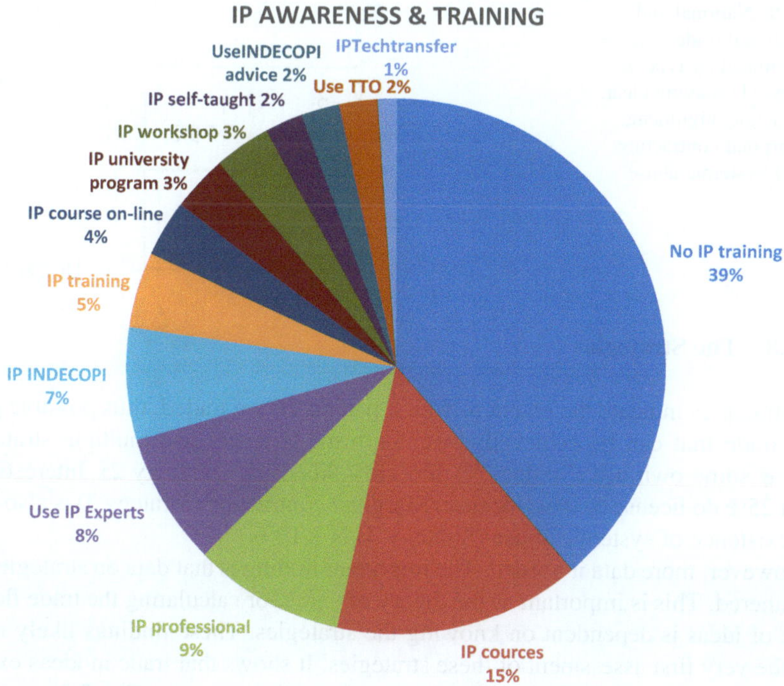

Fig. 7.6 IP awareness

Fig. 7.7 Patent portfolio growth

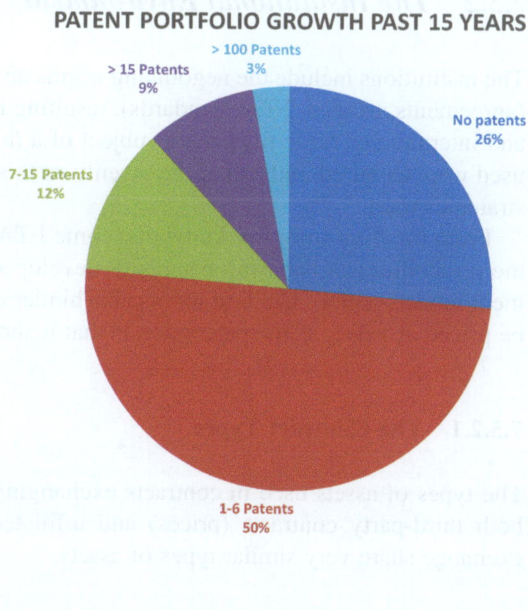

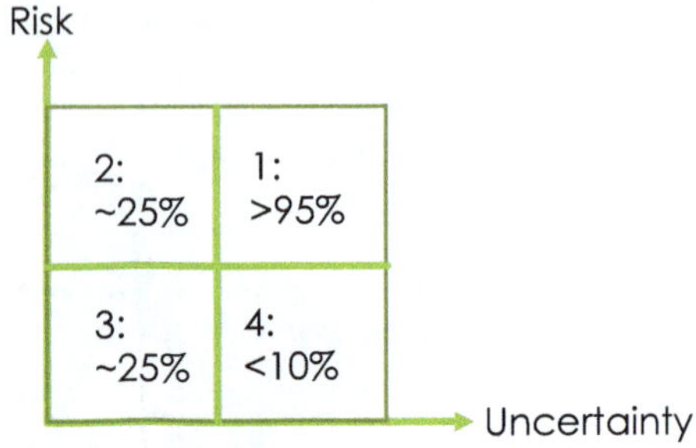

Fig. 7.8 National and international trade in ideas differentiated by type of strategy: (1) Staying clear, (2) Strategic alignment, (3) Marginal contracting, and (4) Systemic abuse

7.5.1.8 The Strategies

The strategies indicate the extent to which patents/IP are traded, thus possible gains from trade that can be achieved. Inventor firms typically use multiple strategies, such as some own use (Strategy 1) and cross-licensing (Strategy 2). Interestingly, about 25% do license or cross-license. Marginal contracting (Strategy 3) is also high. The existence of systemic abuse (Strategy 4) is <10%.

However, more data it needed. The important finding is that data on strategies can be gathered. This is important as the theory of value for calculating the trade flow of value of ideas is dependent on knowing the strategies. These findings likely represent the very first assessment of these strategies. It shows that trade in ideas exist in developing nations (a smaller part is *international* licenses) (see Fig. 7.8).

7.5.2 The Institutional Environment

The institutions include the negotiating norms and rules and international trade rules (agreements, treaties, FTA, standards), resulting in international contracts. The rules and international trade rules are a subject of a *further study*. The types of contracts used were captured, indicating the incentives to share risk and reflecting the type of strategies used.

From the pilot study we know that some FTA and government regulations limit the possibilities of licensing patents/IP developed by universities (whether government funded or not). Cultural norms also hinder university-developed technology to be priced at value, as the perception is that it should be "free."

7.5.2.1 The Contract Types

The types of assets used in contracts exchanging patents. The transactions include both third-party contracts (prices) and affiliated (transfer prices). Both types of exchange share very similar types of assets.

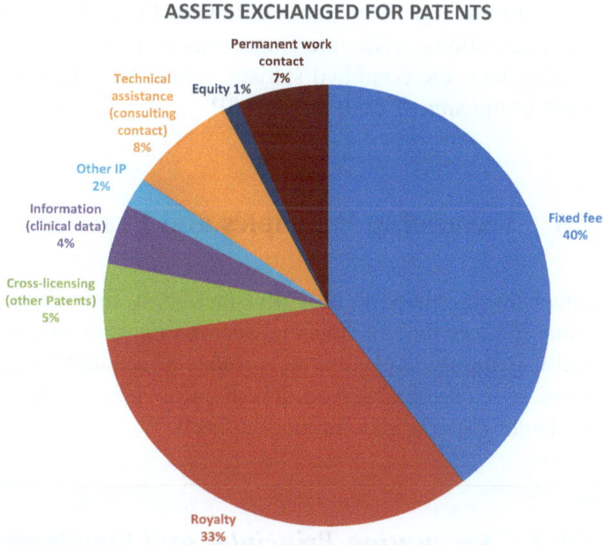

Fig. 7.9 Assets exchanged for patents

It is clear that it is necessary to include both cross-licensing agreements (other patents), other IP, consulting, and equity in the transaction. This means that the balance sheet needs to include accounts on IP assets (not only "Goodwill") in order to properly measure trade flows from patents/IP (see Fig. 7.9).

7.5.3 Additional Data Needed for Statistical Purposes

Institutional rules represent the most important addition. This would allow the value flow from trade in ideas to be calculated given the strategies *and* the institutional trade rules arrangements. (Such a study is planned for the "next phase" of the trade in ideas program, of which this POC was a part.)

7.5.4 The Outcome Measures

The outcome measures such as *Return on IP Assets* (RIPA) and *terms of trade* (export/import prices) by different technology areas can likely be calculated based on data on patents, inventors, and specialized survey. In addition, an international patent position (IPP) can be created as a new statement with these data. Having this data on technology areas (IPC classification) over time allows to calculate gains from specialization in different areas as a result of trade in ideas.

Institutional performance would need the additional study of existing trade norms and international trade rules (agreements, treaties, conventions, standards, etc.).

Together, the compiled statistic would be able to inform major policy development in the area of trade in ideas/IP.

7.6 Accounting Principles and Practices

The answers related to how revenues, cost, financing, and assets are accounted for today indicate that the financial accounting standards used are observed. This means that it is highly likely that new financial accounting standards for IP revenues and assets will also be observed. It is therefore possible to use standards as an instrument for better data on trade in ideas as well.

7.6.1 Accounting Principles and Standards

The financial accounting standards used for recognition of *revenues* are 75% national standards and 25% IFRS15 (with 26 reporting and 6 don't know). It is safe to say that given the size of the companies the national standards prevail. When it comes to assets and liabilities, 60% of patents are expensed and 40% capitalized.

Contracts are rarely capitalized as assets or liabilities (10–20%?). *R&D* is both expensed and capitalized but also rare (20%?).

The number of respondents is low on these items and more data is needed to make any judgement. However, the fact that there is some reporting implies a "beginning" of these kinds of principles, viewing IP as assets. A long road likely lies ahead.

7.6.2 Accounting Practices

7.6.2.1 Revenues

About one-third of respondents differentiate the recognition of revenues by different types of IP, whether sale or licensing, technology area, national and international, and types of other assets in the contract.

7.6.2.2 Cost

Close to 40% allocate cost separately to: the development process (research and development), selling/licensing/enforcing, portfolio management, technology area, and partner contracting such as with universities.

7.6.2.3 Assets and Liabilities

Most do expense patent investment and costs, but some do in fact capitalize and depreciate these over 5–10 year. However, there were too few answers here to draw any definitive conclusion.

7.6.2.4 Funding

A range of funding sources are used from equity, grants, prizes, and loans. They are accounted under financial items or extraordinary items.

7.7 What Can We Learn from the POC?

The bottom line is that it is quite likely that data on key elements to calculate statistics can be collected. This would be through special surveys where respondents can refer to their own financial accounts.

The development of such specialized surveys and new international financial accounting standards (and recommendations) seems the next step in the process of developing measuring standards for trade in ideas.

Chapter 8
Can Data be Collected in a Developing Nation?

The basis element in collecting data on trade in ideas is a *culture* to "honor the inventor." This has been the principle of the patent system since its first conception in Venice in 1474. The purpose at the time was to invite these people to the city state with their ideas, which made the early manufacture industry prosper. This culture of inviting inventors from other countries to come with their ideas has since been a very good economic strategy for many nations. Today, with almost global property rights on ideas, governments thus need to embrace a *trade* policy that gives incentives to the holders of these rights-protected ideas to engage in trade in ideas.

The fundament is thus not only to protect the ideas that people have, but also to protect and develop the markets for these ideas. This puts two conditions on whether a country can collect data: (1) a culture of honoring the inventor and (2) a culture of export (trade)-oriented development policy. Together they create a new economic development policy of North–South trade in ideas.

To the first point, data on the inventor agent and characteristics including their strategies are needed, and to the second point, data on the institutional trade rules are needed.

8.1 An International Effort of Sovereign Nation States

The data on the *economic environment* is mainly a national issue, but the data on patenting, company organization, university research, human capital formation, financial capital, and so on come today from a whole range of agencies. Therefore, there needs to be a coordination of these agencies' statistical work, when it comes to IP and human capital formation. Fortunately, in the proof of concept (POC), Perú had the only(?) agency that spans across all IP issues. All other nations need to coordinate the work in different agencies, either formally or through *special working groups*.

E. Ullberg, *Intellectual Property Statistics*, Contributions to Economics,
https://doi.org/10.1007/978-3-031-36386-3_8

The data on the *institutions* is mainly a poly-lateral and multilateral issue, which thus means that regional economic cooperation areas and the only multilateral agency, i.e., the WTO, need to be involved in connecting the dots on trade rules related to the cross-border transfer and licensing of patents/IP. This is likely such a big task, that one should start at the regional level and then the multilateral levels. As there are hundreds of bilateral and poly-lateral agreements but only a handful of multilateral ones, it may be necessary to first select a few trading regions and then create a more global picture based on the initial findings from the regions. This work could therefore be done through *working groups* as well, at the regional and international levels.

A special working group is needed to coordinate these two perspectives – national and international – so that the national–cultural dimension will be included and an international-standardized dimension can be developed.

This group was created at WTO in 2019, with Perú as the founding member, with the name "The Informal Team of the Trade in Ideas Program." Among its activities has been proposed to build on the developed statistics framework and further develop the necessary addition to standard and statements outlined in this book.

It is necessary to have the national dimension at heart because cultural norms, which structure the exchange along with formal rules, are an *independent variable* in the proposed statistical analysis.

Given that most developing countries are part of WTO, it is possible to combine the cultural and formal rules in this way. This approach of "country-led" workshops make it likely that the data can be collected.

8.2 The Data Needs and Standardization Needs

Data needed is based on financial accounting standards and reporting, patent office standards and registry, and statistical agencies surveys. A coordinated approach on the general theme of regional and international trade can thus create a learning between countries and provide ideas to resolve problems and inspire solutions to cultural–trade economic data collection principles and standards and methods.

Without the cultural dimension, which is so central to the functioning of any society, one simply cannot succeed. As the culture is an "internal process" of the country, this work can take place at national levels. This also goes to the heart of the management of invention problem, where corporate cultures clearly operate within a national culture. However, with multinational corporations and the "minimum standards" approach to international economic law (of which statistics is a part), it is clearly a "cultural trade-off" that is needed on both parts. A preferred approach is therefore *national treatment*. However, national treatment has the "drawback" of not being standardized. This topic was touched upon in the introduction (Chap. 1) the statistics chapter (Chap. 5). The treatment of multinational companies is an important one, as they are the largest holders of patents. However, the small- and medium-sized firms and universities are the most inventive agents. The role of international business and international researchers and inventors need to be further examined to keep a sovereign nation state system as the basis of world order. This has proven far superior to regional hegemonies since its conception and is the basis for a level playing field for (economically) smaller nations, which is the definition of developing nations.

Without the international dimension, which is central to standardization, one cannot create comparable statistics, so this discussion of national–culture and international is needed.

8.3 An International Discussion Needed

The conclusion is that it is necessary to have an initiative based on the sovereign nation states where free states, free businesses (and other agents), and free inventors can access the international market with their ideas through trade.

The proposal is thus to structure the work with the standardization of data collection through an international working group represented by sovereign member states that are also part of a multilateral trading system (currently WTO) and to create national (and regional) working groups at the national level to coordinate the work of different agencies. In all cases the key agent in this market, the *inventors* (individuals, teams, SMEs, MNCs, university–institutes), need to be represented together with the policymakers in charge. This approach allows data to be collected from developing nations as well as developed nations to create standards and measures of trade in ideas.

In the end, such a scenario may then, when the data is used to inform trade policy, transform the WTO to an institution for the twenty-first century's idea-based economic system. See earlier reference discussion by Phelps and Ullberg (2018).

8.4 The Primary Tools

A primary tool would be *specialized surveys* sent out by the patent offices of a ministry of trade agency. This is likely necessary given the disparate and mostly uncoordinated situation at national levels and data quality issues at the international levels.

For example, during the pilot study, no data received from WIPO could be used to identify the inventors as the quality was too poor. For example, many patents from Sweden, coded SE in the UN system, had been coded with El Salvador's code, SV. As a result, in every case the national offices had to get involved and search their databases, in some cases manually, spending weeks to collect the actual inventor information. This is surprising as the data consolidated by WIPO come directly from the patent offices. These may be exceptions, but they could be causes for concern.

For a very long time, inventors were also coded by their *name* in the patent systems (although corporate registries exist in most countries), resulting in sometimes hundreds of different spellings of the same company. The update of ownership is also lacking, leaving ownership, i.e., who owns the production asset, in the dark. The reason for this is likely that the patent offices are organized around the patents, and they "gran, challenge, and enforce patents," neither the *inventor* nor the economic *legal entities* holding them. The economic dimension is not yet integrated.[1] Private companies have since long provided better data quality.

The connection to the legal entity identification is also necessary to make, as this is related to the international trade transactions. Surveys are therefore an important tool, probably for a long time, until more electronic methods to capture the data at the source – the companies, patent offices, official registries, and central banks.

It must therefore be a national initiative at its heart with international cooperation between the inventors and governments to be successful as the data comes from so many different sources.

[1] "As a matter of curiosity, an early attempt to deal with this question was initiated by EPO in the early 2000s. See internal report of the EPO by McGinley and Ullberg, 2002. The result of this work was that the first chief economist was hired to any major patent office. He then started his work by making the patent information available for research. This initiative became "fashionable" and today all major patent and IP offices have chief economists." The report's content was also presented at several public events from 2002–2006 organized by EPO and others, starting at the first Annual conference of *epoline* in Nice, France, Dec 3–5, 2002. "The internal EPO report is referenced on p. 3 in: Gianotti P., "Value, cost and volatility along the lifetime of patent rights: risk management perspectives on the patenting process", 4th EPIP Conference - October 1 and 2, 2004, University of Paris Dauphine, Paris - France. This article is, in turn, based on the "risk management" approach to patents adopted by EPO based on the research referenced earlier. See (Ullberg et al., 2002)".

Part V
Implementing the Framework

Chapter 9
Future Steps Toward New Standards and Statistics

This book should be seen as an input to the discussion of future international balance of payment statistics, international financial accounting standards, and other standards related to patents and trade in ideas (IP).

9.1 The Way Forward – BPM7, IFRS15, IAS38, and Standardization Bodies

Currently, there is a review of the Balance of Payment Manual, Version 6 (BPM6). A presentation of the framework and implication for BPM would be evaluated with the relevant committees: would that be interesting? This could be done through a white paper outlining the IP statistic framework based on this book.

The most important international agency is certainly IMF, which is responsible for the BPM7 update.

Other agencies are then international financial accounting organizations. The most important are IAS with the IFRS15 standard and USGAAP.

These efforts need to focus on the new approach to IP assets and liabilities. They also need to focus on a formal business and economics-based definition of "products," including ideas, goods, and services. The proposal put forth is that the logical structure comes from the *provenance*, i.e., which production assets are used to produce the products: intellectual property, fixed assets, or human capital.

E. Ullberg, *Intellectual Property Statistics*, Contributions to Economics,
https://doi.org/10.1007/978-3-031-36386-3_9

9.2 Process for Adjusting the Standards

This process includes both an *internal process* – expanding data collection based on extant standards and regional field experiments aimed at learning more about the management of inventions in different cultural settings – and an *external process* – expanding the standards in international cooperation.

A first step would be to increase the *special schedule* for data collection that is already part of BPM6. This can be expanded to capture some of the most important patent/IP-related items. The data can then be used to create a balance of payment statistics after the structure presented in this book. This would give a first test of the statistics that could be created.

A second step would be to enter a dialogue with the IMF's standards bodies. Such a dialogue is at the national agency level, and it would thus be initiated and coordinated by the proposed working groups.

A third step could be to engage in the actual standardization process with IMF, through the interested governments. Such a process would need real data from the national level on inventors. To get this data, a process of *regional* field experiments is proposed where countries in the same economic cooperation area make full-scale tests to value the trade flows from patents/IP in some key technology areas or markets. This would yield important insights where the cultural dimensions would also be taken into consideration.

Other standards follow other procedures but all related standards, especially the international financial accounting standards processes, should be an entry point to cooperation: would that be of interest to these bodies?

Again, national and regional tests through field experiments at regional levels would be the approach forward to learn how the *management of inventions* is really carried out businesses and universities–institutes and what that means for the accounting needs related to the creation, use, and trade of patent/IP. The culturally related management issues here are of special importance.

Additional and related standards should be listed so that a better overall system of data can be achieved while coordinating national and international agencies and standards.

Finally, new statistical and accounting statements are needed to inform trade policy. The most important may be return on IP assets (RIPA), terms of trade per tech area, and an international patent position (IPP), as well as a national index on trade in ideas (IP) (see Chap. 2). This would include the formal codification of strategies firms use to evaluate institutional performance. These statements will, for the countries involved, better reflect the risk and uncertainty in the management of inventions and the trade in ideas.

Bibliography

Allais, M. (1951). *Notes théoriques sur l'incertitude de l'avenir et le risque*. Source: Fonds d'archives privées de Maurice Allais (The Maurice Allais Foundations Private Archives).

Allais, M. (1952). Generalisation des théories de l'équilibre économique général et du rendement social au cas du risque. In *Communication au Colloque International sur le Risque* (pp. 1–20), Colloque International de Paris, Actes du Colloque, pp. 81–120

Allais, M. (1953). L'Extension des théories de l'équilibre économique générale et du rendement social au cas du risque. *Econometrica, 21*, 269–290.

Arrow, K. J. (1964). The role of securities in the optimal allocation of risk-bearing. *The Review of Economic Studies, 31*, 91–96.

Aumann, R. J., & Dreze, J. H. (2009). Assessing Strategic Risk. *American Economic Journal: Microeconomics, 1*(1), 1–16. https://doi.org/10.1257/mic.1.1.1

Coase, R. H. (1960). The problem of social cost. *Journal of Law and Economics, 3*, 1–44.

Debreu, G. (1959). *Theory of value*. Yale University Press.

Ericsson Annual Report 2021. (2021).

Hart, O. D. (1988). Incomplete contracts and the theory of the firm. *Journal of Law, Economics, & Organization, 4*, 119–139.

Holt, C. A., Jr., & Roth, A. E. (2004). The Nash equilibrium: A perspective. *Proceedings of the National Academy of Sciences of the United States of America, 101*, 3999–4002.

Knight, F. H. (1921). *Risk, uncertainty and profit*. AM Kelley.

Lev, B. (2018). Ending the accounting-for-intangibles status quo. *European Accounting Review, 28*, 1–24. https://doi.org/10.1080/09638180.2018.1521614

McGinley, C., & Ullberg, E. (2002). Creating an efficient market in ideas. Internal report for the European Patent Office.

Nash, J. F. (1950). The bargaining problem. *Econometrica, 18*, 155–162.

North, D. C. (1981). *Structure and change in economic history*. W.W. Norton & Company, Inc.

North, D. C. (1991). Institutions. *The Journal of Economic Perspectives, 5*, 97–112.

Paton, W. A., & Littleton, A. C. (1940). *An introduction to corporate accounting standards*. American Accounting Association.

Phelps, E. S. (2019). *The three revolutions economics needs | by Edmund S. Phelps* [WWW Document]. Project Syndicate. URL https://www.project-syndicate.org/commentary/economics-must-change-in-three-ways-by-edmund-s%2D%2Dphelps-2019-01. Accessed July 13, 2019.

E. Ullberg, *Intellectual Property Statistics*, Contributions to Economics,
https://doi.org/10.1007/978-3-031-36386-3

Phelps, E. S., & Ullberg, E. (2018). *The need for a global patent market | by Edmund S. Phelps & Eskil Ullberg* [WWW Document]. Project Syndicate. URL https://www.project-syndicate.org/commentary/integrated-global-patent-market-by-edmund-s%2D%2Dphelps-and-eskil-ullberg-2018-08. Accessed August 10, 2018.

Qualcomm Annual Report 2022. (2022).

Razgaitis, R. (2004). U.S./Canadian licensing in 2003: Survey results. *Les Nouvelles, 39*, 139–151.

Robinson, J. (1977). What are the questions? *Journal of Economic Literature, 15*, 1318–1339.

Schumpeter, J. (1912). *Die Theorie der wirtschaftlichen Entwicklung*. Duncker & Humbolt.

Schumpeter, J. A. (1934). The fundamental phenomenon of economic development. In J. A. Schumpeter (Ed.), *The theory of economic development* (pp. 57–94). Harvard University Press.

Smith, V. L. (1973). *Notes on some literature in experimental economics* (Social Science Working Paper No. 21). California Institute of Technology, pp. 1–27.

Smith, V. L. (1982). Microeconomic systems as an experimental science. *The American Economic Review, 72*, 923–955.

Ullberg, E. (2009). *From personal to impersonal exchange in ideas – Experimental study of trade in organized markets for patents*. KTH TRITA-TEC-PHD 09-006, 180.

Ullberg, E. (2012). *Trade in ideas: Performance and behavioral properties of markets in patents*. Springer.

Ullberg, E. (2015a). *The language of trust and reciprocity in markets in patents: A sociological analysis of property rights on messages resolving uncertainty in exchange in ideas* (IP2 Working Paper No. 15016). Stanford IP2 Program.

Ullberg, E. (Ed.). (2015b). *New perspectives on internationalization and competitiveness*. Springer International Publishing.

Ullberg, E. (2016). Coordination of inventions and innovations through patent markets with prices. *Journal of the Knowledge Economy, 8*, 704–738. https://doi.org/10.1007/s13132-016-0418-0

Ullberg, E. (2018). *What is the economic potential of trade in ideas?* (Note on The trade in ideas program: Next steps in program development, WTO Nov 26, 2018).

Ullberg, E. (2019). Economic efficiency and field-of-use pricing of SEP licences under FRAND terms. *Queen Mary Journal of Intellectual Property, 9*, 392–413. https://doi.org/10.4337/qmjip.2019.04.02

Ullberg, E., Rodriguez, E., & Stormby, N. (2002). Risk management: From portfolio strategy to value–creating systems strategy. *Geneva Papers on Risk and Insurance – Issues and Practice, 27*, 467–476. https://doi.org/10.1111/1468-0440.00185

Author Index

A
Allais, M., 29
Arrow, K.J., 29
Aumann, R., 40

C
Coase, R.H., 95, 108
Cray, S., 9

D
Debreu, G., 29

E
Einstein, A., 30

H
Hart, O.D., 95
Hawkins, S., 30
Holt, C.A. Jr., 40

K
Keynes, J.M., 29
Knight, F.H., 29, 40

L
Lev, B., 71, 94
Littleton, A.C., 71

M
Magdeleine, J., 53
Maurer, A., 53

N
Nash, J.F., 39, 40
North, D.C., 7, 11

P
Pato, W.A., 71
Phelps, E.S., 15, 29, 169

R
Razgaitis, R., 67
Robinson, J., 44
Roth, R.E., 40

S
Shakespeare, 77
Smith, A., 27, 33
Smith, V.L., 8, 9, 18, 31, 39, 41

U
Ullberg, E., x, 6–8, 13, 15, 35, 38, 39, 53, 72, 78, 93, 108, 116, 128, 160, 169

E. Ullberg, *Intellectual Property Statistics*, Contributions to Economics,
https://doi.org/10.1007/978-3-031-36386-3

Subject Index

A

Accounting
- assets, 17, 19, 30, 40, 50, 71, 75, 78–80, 82, 85, 86, 88–90, 92, 98, 102, 104, 109, 112, 113, 120, 129, 133, 164, 173
- balance sheet, 17, 22, 73, 79, 82, 90–92, 97, 98
- contract, 9, 17–19, 21, 41, 46, 73, 79, 87, 90, 92, 96–98, 112, 119, 120, 127, 133, 156, 157, 164
- income statement, 17, 22, 73–79, 96, 97
- IP as assets, 40, 78, 80, 93, 104, 164
- liabilities, 17, 74, 78, 80–82, 88, 90–96, 98, 104–105, 109, 120, 129, 164, 173
- licensing contracts as assets, 74, 91
- licensing contracts as liabilities, 17, 81, 82, 91, 92, 96, 105, 120, 130, 164
- patent family, 124, 154
- patents as assets, 36
- principles, 7, 17, 19, 85, 120, 157, 164–165
- product, 91, 93, 119
- provenance, 94, 96
- revenues, 19, 73–79, 87, 89, 92–94, 96, 104, 113, 120, 153, 157, 164
- standards, 4, 7, 9, 16, 19, 30, 40, 41, 69, 71–98, 102, 104, 112–114, 120, 156, 157, 164, 165, 168, 173, 174

Agent
- business model, 107, 136
- characteristics, 9, 11, 31, 33, 34, 43, 46, 106, 107, 114, 117, 121, 136, 153, 157
- companies, 17, 43, 106, 116, 129, 168
- hierarchy, 12, 18, 60, 61
- innovator, 32
- inventor, 12, 26, 31, 32, 34, 44, 47–50, 106, 116, 129, 151, 153, 157, 167, 169
- network, 64, 87
- standardization organizations, 168, 169
- states, 169
- trader, 32
- universitites, 11

Assets
- chose in action, 67, 74, 75, 96–98
- chose in possession, 67, 74, 75, 79, 88–90, 96–98
- contracts as assets, 105
- intangible assets, 22, 27, 71, 74, 79–90, 92, 97, 98, 118, 144
- intellectual property right (IPR), 4, 67, 90–92, 94, 95, 101, 103, 104, 134
- patent asset, 39, 46
- patent family, 154
- tangible assets, 83, 86, 92, 94, 95, 103, 120, 130
- thing in action, 74, 128
- thing in possession, 74, 128

B

Balance of payment (BOP)
- anonymized data, 136
- BPM5, 68, 120, 133–136, 138–147
- BPM6, 7, 21, 54, 55, 58, 68, 120, 127, 130, 173
- BPM7, 120, 130, 173
- capital account, 21, 129, 130, 135, 141–143
- change in principle, 93
- current account, 21, 93, 104, 129, 130

E. Ullberg, *Intellectual Property Statistics*, Contributions to Economics, https://doi.org/10.1007/978-3-031-36386-3

Balance of payment (BOP) (*cont.*)
current account in ideas, 21, 93
current account in goods, 93
current account in services, 93
data collection, 55, 60, 64, 96, 120–125, 127, 174
financial account, ix, 21, 30, 93, 104, 111, 113, 120, 127, 129, 173
logic, 133
statistical discrepancy, 21
trade in goods account, 6, 7, 21, 93, 130, 142
trade in ideas account, ix, x, 7, 21, 22, 53, 55, 60, 63, 66, 67, 69, 74, 93, 99, 104, 113, 114, 120, 122–124, 130, 131, 133–136, 138–145, 151, 152, 163, 164
trade in services account, 6, 7, 21, 54, 58, 61, 63, 65, 104, 130, 139, 142
Behavioral properties, 3, 10, 11, 13, 42, 45, 102, 109, 110
BOP, *see* Balance of payment (BOP)

C
Changes in principles, 99, 127–131
Conceptual framework
accounting, 129, 130
culture, 167
economic, 55–57, 167, 169
international economic law, 12, 18, 25, 30, 32, 35, 61, 168
statistics, 55, 64
Contracts
contract as assets, 12, 17, 19, 80, 90, 96, 103, 105
contracts on patents, 5, 11, 22, 26, 73, 77, 87
fixed fee, 25, 41, 79, 90, 93, 94, 97, 98, 111, 119, 131, 141, 145, 154
incentive, 16, 106, 162
incomplete, 12, 23, 95, 101, 105, 107, 111, 119, 124, 131, 140
on patents, 17, 22, 117, 120, 122, 127, 156
royalty, 25, 26, 41, 47, 78, 79, 90, 91, 94, 97, 111, 119, 124, 131, 139, 141, 142, 154
two-part tariff, 25, 26, 111, 119, 123
Coordination, 29, 61, 167
Creative company, 23, 72
Creativity
creative company, 23, 72
culture of, 8–9
human, ix
Culture of creativity, 8–9

D
Data
data collection structure, 127–131
data gap in financial accounting, 71–98
data gap in statistics, 53–69, 96
Data gap
balance of payment, 53–69
financial accounting, 4, 71–98
statistics, 53–69, 96
Dynamic
economic system, ix, 4, 13, 25, 27–35, 49, 66–69, 101, 103, 106, 114
economic value, 112–113

E
Economic
agent characteristics, 9, 11, 31, 34, 46, 106, 107, 114, 117, 121, 153, 157
agents, 9, 11, 12, 26, 31, 34, 47, 102, 106, 114, 116, 121, 129, 153, 157
environment, 9, 11–12, 14, 16, 19, 20, 26, 31, 32, 41, 46–48, 101–112, 121, 136, 153, 157–162, 167
growth, ix, 101
institution, 3, 9, 13, 14, 16, 19, 26, 42, 45, 62, 106, 110, 114, 123, 153, 157, 168
risk, 4, 12, 14, 16, 25–29, 34, 40, 62, 80, 106, 108, 110, 118
system, 3, 25, 56, 72, 101, 133, 151, 169
uncertainty, 4, 12, 14, 16, 19, 25–31, 34–42, 45–47, 62, 75, 78–80, 83, 85, 89, 94, 106–108, 110, 117, 118, 130, 174
Empathic management, 72, 73
Exchange, 3, 5, 6, 9, 10, 14–18, 26, 27, 29–36, 38, 39, 42, 43, 48, 75, 79–83, 86, 88, 93, 97, 102, 106, 109–111, 123, 128, 131, 133, 135, 139, 141, 144, 145, 152, 153, 162, 168

F
Financial accounting
balance sheet, 17, 40, 73, 78, 79, 90, 93, 96–98, 153
contracts, 16–20, 30, 41, 46, 50, 73–76, 78–80, 90, 96–98, 120, 127, 153, 156, 164
development phase, 79, 80, 85–86, 97, 142
income statement, 17, 22, 40, 73, 74, 79, 80, 96, 97, 153
international financial accounting, 7, 16, 19, 21, 30, 72–74, 114, 165, 173, 174

IP as assets, 40, 78, 80, 93, 103, 104, 127, 164
IP contracts as assets, 104
IP contracts as liabilities, 22
licensing contracts as assets, 105
licensing contracts as liabilities, 105
patent family, 124, 154
patents as assets, 30
recognition of assets from IP, 129, 130
recognition of cost from IP, 79, 80, 83–84, 97
recognition of IP as assets, 92, 93
recognition of revenues, 19, 73, 74, 76, 79, 93, 94, 96, 164
recognition of revenues from IP, 92–94
research phase, 79, 80, 85, 89, 97
revaluation, 79, 82, 88, 90, 92, 97, 98
revenues, 19, 20, 74–79, 96, 97, 104, 113, 120, 129, 153, 164
standards, 4, 7, 16, 19, 30, 40, 41, 69, 71–98, 113, 114, 120, 156, 164, 168, 173, 174
taxes, 17, 47, 79, 97, 153, 156
Financial analysis
COGS, 22
EAT, 79, 97, 143
EBIT, 22, 79, 97, 143
EBT, 79, 97, 143
HCF (*see* Human capital formation (HFC))
IFRS, 73, 75, 78, 80, 81, 84, 88, 120, 142, 144
IPP (*see* International Patent Position (IPP)
patent position, 120, 130, 145
revenues, 22, 39, 75, 76, 79, 94–95, 103, 130
RIPA, 21, 50, 73, 80, 82, 95, 104, 113, 147, 163, 174
Terms of trade, 16, 21, 50, 74, 104, 113, 151, 163, 174

G
Gains from specialization, 5, 6, 12, 14–16, 21, 32, 34, 38, 42, 74, 113

H
Human capital formation (HFC), ix, 8, 11, 13, 16, 23–27, 38, 44, 47, 72, 84, 87, 103, 104, 117, 122, 131, 141, 151, 157, 159, 160, 167
Human creativity, ix
Human ideas, 5, 9, 28, 29, 77

I
Idea, 3, 25, 53, 72, 101, 127, 151, 167, 173
Idea-based economic system, 7, 19, 57, 60, 66, 67, 102
Impersonal exchange, 5, 41
Industrial organization, 32, 43, 67
Information
institutional environment, 33, 49
institutional learning, 49
institutional performance, 42
institutional policy, 9
institutions, 25, 26, 33, 41, 49, 131, 157
Institutional performance, 3, 10, 11, 13, 42, 45, 72, 86, 102, 109, 110, 130, 164, 174
Integration
integration of science and technology, 7–8, 12, 30, 44, 47, 143, 146
international markets, 4, 108
Intellectual property (IP)
chose in action, 67, 74, 75, 91, 97, 98
chose in possession, 67, 68, 74, 75, 79, 88, 89, 91, 96, 98
embedded IP, 27, 53–55, 66, 77, 81, 83, 88
embedded, 27, 53–56, 66, 71, 77, 81, 83, 88
embodied, 54, 58, 122, 141, 142
IP assets, 4, 40, 55, 73, 103, 129, 163, 173, 174
IPP, 22, 61–63, 65, 104, 113, 120, 125, 130, 131, 133, 134, 143–147, 163, 174
IP traders, 64–65
non-embedded (*see* Intellectual property (IP), non-embedded IP)
non-embedded IP, 27, 55, 77
patents, 3–23, 25–27, 32–35, 40, 44, 49, 53–57, 59–62, 66, 71, 72, 74–78, 80, 81, 84–98, 101, 103–111, 113, 114, 116, 117, 119–122, 124, 125, 127–133, 140, 142–145, 153, 155, 160, 163, 173
Intellectual property rights, *see* Intellectual property (IP)
International Financial Accounting Standards
IAS38, 73, 74, 76, 79–82, 85, 87, 90, 97, 103, 124, 130, 133–136, 138–147
IFRS, 73, 75–78, 80–82, 84, 88, 120, 127, 142, 144
IFRS15, 19, 73–76, 78, 79, 87, 90, 93, 96, 97, 103, 120–124, 130, 134–136, 138–147, 164, 173
US GAAP, 73, 78, 173
International organizations
ECOSOC, xxi
IMF, 173, 174
OECD, 8, 55, 59, 65, 125

International organizations (*cont.*)
WTO, 13–15, 48, 53, 67, 110, 168, 169
International Patent Position (IPP), 22, 55, 56, 61–63, 92, 104, 113, 130, 163, 174
International trade
policy, 32, 151
rules, 10, 15, 86, 123, 130, 151, 162, 164
Inventor, 3–9, 12–19, 23, 26, 28, 30–32, 34, 36, 38, 39, 41, 42, 44, 45, 47–50, 61, 78, 81, 88, 95, 96, 105, 106, 108–111, 114, 116, 118, 119, 129, 131, 133, 135, 151–154, 156–160, 162, 163, 167, 169, 170, 174
Investment strategy, 31, 32, 40–43, 47, 73, 78, 80, 107, 109, 123
IP, *see* Intellectual property (IP)
IP assets, 4, 8, 11, 21, 40, 55, 66, 73, 80, 82, 84–86, 92, 94–96, 104–105, 109, 112, 116, 129, 173

K

Key concepts
anonymized data, 17
anonymous, 17 (*see also* Private)
current account for trade in ideas, 21, 104, 130, 135
embodied (*see* Intellectual property, embodied)
embedded (*see* Intellectual property, embedded)
idea economy, 6, 12
IP as assets, 40, 78, 80, 93, 103, 104, 127, 164
IPP (*see* International Patent Position (IPP)
legal entity, 16, 116, 135, 158–159
licensing contracts as assets, 14
licensing contracts as liabilities, 17, 81, 82, 91, 92, 96, 105, 120, 130, 164
patents as assets, 36
provenance, 94–96, 103–104, 130, 173
reward structure, 37–38
reward structure (formula 2.4), 38
RIPA (*see* Financial analysis, RIPA)
statistical unit, 13, 114–119, 129, 135
strategies which resolve uncertainty (table 2.1), 36–37
terms of trade in patented technology, 3, 10, 12, 13, 15, 16, 22, 28, 36, 38, 41, 75, 93, 122, 143, 152
transformation, 10

L

Learning, 12–14, 16, 19, 27, 40, 44, 46–49, 72, 121, 133, 168, 174
Legal entities
creative company, 116, 158, 159, 170
statistical unit, 16, 114, 116, 117, 120, 121, 129, 135, 158
Level playing field, 13, 18, 50
Liabilities, 17, 21, 22, 47, 49, 62, 74, 78, 80–88, 90–96, 98, 103–105, 109, 118, 120, 122, 127, 129–132, 134, 139, 144, 145, 156, 164, 165, 173
License, 6, 25, 56, 75, 103, 127, 153
Logic, 103, 116, 127, 128, 133

M

Management
empathic, 72, 73
of inventions, 22, 72, 73, 76, 79, 80, 82, 88–90, 96–98, 102, 118, 129, 133–136, 138–147, 157, 158, 160, 168, 174
of inventors, 8, 30, 41, 42, 129, 157, 158, 160, 168
of talent, 72, 73, 158
Marginal contracting, 36–38, 42, 43, 108, 109, 118, 162
Marginal contracts, 42
Markets in patents, 3, 4, 9, 10, 12, 13, 16, 18, 45, 102, 110, 131, 143, 156
Measures, 10, 25, 53, 71, 101, 127, 152, 169
Measuring, 3, 4, 7, 9, 11, 13, 16–19, 23, 55, 62–64, 74, 80–89, 95, 103, 104, 112, 114, 165
Measuring framework, 7, 114
Messages, 9, 29, 31, 48, 106–108
Mutually assured destruction (MAD), 36
Mutually assured self-restraint (MUST), 35, 37, 42

N

New statistical framework, *see* Statistical framework
Non-satiation postulate (formula 2.5), 42
Non-satiation postulate for uncertain values, 41–42

O
Ownership, 10, 17, 21, 31, 33, 34, 36, 43, 44, 47, 50, 55, 58, 60–63, 65, 66, 69, 84, 92, 96, 103–105, 109, 111, 112, 116–118, 120, 121, 124, 125, 130, 136, 140, 146, 147, 156, 158, 159, 170

P
Patent
  licence, 23, 33, 57, 77, 154
  patents, 3, 25, 53, 71, 101, 127, 151, 167, 173
  sale, 11, 22, 33, 44, 90
  transfer, 4, 5, 8, 9, 11, 14, 16, 18, 21, 31, 32, 34, 38, 39, 41, 59, 69, 75, 76, 94, 117–119, 122, 168
Patent family, 5, 124, 154
Patent system, 5, 6, 10, 11, 13–15, 18, 32–34, 38, 45, 48, 49, 107, 110, 129, 154, 157, 167, 170
Pilot-study, x, 8, 10, 153, 160, 162, 169
Policy, 3, 4, 6–10, 12–15, 19, 35, 36, 42, 44–48, 54, 55, 61, 65, 67, 82, 110, 111, 113, 123, 130, 152, 164
Price (learning sustained growth) Equation 2.3, 46
Price (market clearing conditions) Equation 2.2, 46
Principles
  changes in, 127–131
  of data collection, 96, 120, 127–135, 168
  economic logic, 103
Private, 10–12, 15–17, 20, 26, 31, 32, 37, 38, 48, 50, 61, 62, 67, 69, 87, 107, 108, 111, 116, 122, 127, 136, 137, 139, 157–160, 170
Product, 18, 27, 54, 73, 103, 128, 157, 173
Production assets, 4, 17, 22, 30, 32, 57, 75, 82, 85, 88, 92, 95, 103–105, 117, 118, 120–122, 128–131, 133–135, 144, 170, 173
Productivity, ix, 4–6, 19, 27, 33, 72
Productivity growth, 8, 101, 128
Proof-of-concept (POC), 4, 11, 151–165, 167
Property rights
  on communication, 9, 10, 14, 33, 139
  on exchange, 5, 9, 14, 31
  on information, 94
  on messages, 9, 31
  on things, 75, 103, 134, 139
Provenance
  goods, 94–96, 103, 130, 133, 173
  ideas, 94–96, 103, 130, 133, 173
  of revenues, 94–95, 130
  services, 94–96, 103, 130, 133, 173

R
Recognition of assets from IP, 92, 93, 129, 130
Recognition of cost from IP, 79, 80, 83, 97
Recognition of revenues from IP, 92–94
Research phase, 79, 80, 85, 89, 97, 142
Return on IP assets (RIPA), 21, 50, 73, 80, 82, 95–98, 104, 113, 147, 163, 174
Revaluation, 79, 82, 88, 90, 92, 97, 98, 105, 134, 144
Revenue types
  goods, 76, 79, 89, 94–96, 103, 104, 107, 120–125, 130
  idea, 94, 96, 103, 107, 120–125, 130
  services, 76, 79, 89, 94–96, 103, 104, 107, 120–125, 130
Reward structure, 37–38, 42, 43
Rights, 3, 25, 54, 71, 101, 127, 153, 167
RIPA, *see* Return on IP Assets (RIPA)
Risk, ix, 4, 12, 14, 15, 25–30, 34–37, 39–41, 43, 46, 48, 62, 73, 77–80, 82–85, 94, 97, 106–110, 114, 117, 118, 130, 131, 139, 141, 142, 162, 174
Rules of trade, 10, 106, 107, 110–111, 123

S
Science, 5, 7, 44
Science and technology, 5, 7–8, 12, 30, 44, 47, 49, 50, 143, 146, 157
Scope of data collection, 27, 114
Search costs, 48, 107
Service, 3, 25, 53, 72, 101, 128, 173
Socially-strategic equilibrium, 39–41
Specialization, 5, 6, 12, 14, 15, 21, 25, 32–34, 38, 39, 41, 73, 74, 106–111, 113, 125, 146, 163
Specialization in ideas, 6, 101, 128
Specialization in technology, 3, 12
Standardization, 168–169, 173, 174
Standards, 3, 25, 56, 71, 102, 127, 156, 168, 173
Statistical framework, 7, 14, 16, 53, 54, 66, 69, 72, 101, 102, 114, 151, 152, 168
Statistical principles, 17, 20, 21, 103, 113, 120
Statistical standards, 62, 73
Statistical unit, 13, 16, 114–121, 129, 132, 135, 158

Statistics
agreements, 62
anonymized data, 17
Statistics analysis
international patent position (IPP), 22, 61–63, 65, 104, 113, 120, 125, 129–131, 133, 134, 144–147, 163, 174
technological specialization, 130
terms of trade, 16, 20, 21, 50, 67, 74, 104, 113, 125, 146, 151, 163, 174
Statistics framework, 7, 53, 66, 69, 72, 102, 114, 152, 168
Staying clear, 36–38, 43, 108, 109, 118, 123, 162
Strategic alignment, 36–38, 42, 43, 108, 109, 118, 123, 155, 162
Strategic behavior
behavioral properties, 45, 109
cooperative strategy, 45
marginal contracting, 36–38, 42, 43, 108, 109, 118, 162
non-cooperativ strategy, 16, 40
staying clear, 36–38, 43, 108, 109, 118, 123, 162
strategic alignment, 36–38, 42, 43, 108, 109, 118, 123, 155, 162
systemic abuse, 37, 38, 43, 108, 109, 118, 155, 162
Strategy, 4, 25, 60, 71, 101, 129, 153, 167, 174
Strategy types (formula 2.3), 37
Systemic abuse, 37, 38, 43, 108, 109, 118, 123, 162

T

Tangible assets, 83, 86, 92, 94, 95, 103, 120, 130
Taxes
tax optimization, 12, 17, 19, 50, 54, 55, 61, 112, 116, 136
tax policy, 12, 112
Technology, 3, 27, 60, 71, 101, 130, 151, 174
Terms of trade, 16, 20, 21, 50, 67, 74, 104, 113, 125, 146, 151, 163, 174
Theory of value
choice of strategy, 18, 30
formulas (ref. 2.2), 38
value, 18–19, 25–50, 66, 69, 80, 94–98, 112, 128, 162
willingness to accept, 18, 26, 37, 45, 46, 49, 112
willingness to pay, 18, 37, 45, 46, 112
Things in action, *see* Chose in possession
Things in possession, *see* Chose in possession
Trade economics
comparable advantage, 72
comparative advantage (*see* Trade economics, comparable advantage)
competitive advantage, 72
economics, 17, 25, 50, 168
specialization, 16
terms of trade, 16, 20, 21, 50, 67, 74, 104, 113, 125, 146, 151
Trade in ideas
capital account, 21, 56–58, 129, 130
concept, 93
current account, 21, 73, 93, 102, 104, 129, 130, 132, 135, 141–145
current account in ideas (rights), 130
program, x, 8, 163, 168
trade policy, 4, 13, 20, 26, 32, 45, 48, 50, 65, 67, 86, 133–147, 151, 167, 169, 174
Trade policy, 4, 13, 20, 26, 32, 45, 48, 50, 65, 67, 86, 133–147, 151, 167, 169, 174
Trade secrets, 17, 47, 54, 67, 74, 109, 154
Trade statistics, 50, 65, 114, 136, 138–147
Traditional knowledge, 33, 34, 48, 87, 155
TRIPS, 10, 13–15, 18, 20, 25, 33, 48, 53, 110, 130, 134, 139
Trust, 39, 67, 78, 80, 84, 85, 106
Trust in each other's actions, 36, 37, 39, 83, 123

U

Uncertainty, 4, 12, 14–16, 19, 25–31, 34–37, 39–42, 46–49, 60, 62, 73, 75, 78–80, 83, 85, 89, 94, 106, 108, 110, 117, 118, 130, 162, 174
US GAAP, 73, 78, 173

V

Value, 4, 25, 54, 71, 101, 128, 158, 174

W

WTO, 11, 13–15, 45, 48, 53, 67, 110, 168, 169

MIX
Papier aus verantwortungsvollen Quellen
Paper from responsible sources
FSC® C105338

If you have any concerns about our products,
you can contact us on
ProductSafety@springernature.com

In case Publisher is established outside the EU,
the EU authorized representative is:
Springer Nature Customer Service Center GmbH
Europaplatz 3, 69115 Heidelberg, Germany

Printed by Libri Plureos GmbH
in Hamburg, Germany